THE ENVIRONMENTAL TURN IN POSTWAR SWEDEN

A new history of knowledge

The environmental turn in postwar Sweden

A new history of knowledge

DAVID LARSSON HEIDENBLAD

Translation: Arabella Childs

Lund University Press

Copyright © David Larsson Heidenblad 2021

The right of David Larsson Heidenblad to be identified as the author of this work has been asserted by him in accordance with the Copyright, Designs and Patents Act 1988.

Lund University Press
The Joint Faculties of Humanities and Theology

LUND
UNIVERSITY
PRESS

PO Box 191
SE-221 00 LUND
Sweden
http://lunduniversitypress.lu.se

Lund University Press books are published in collaboration with Manchester University Press.

British Library Cataloguing-in-Publication Data
A catalogue record for this book is available from the British Library

Lund University Press gratefully acknowledges publication assistance from the Thora Ohlsson Foundation (*Thora Ohlssons stiftelse*)

ISBN 978-91-985577-3-2 hardback

ISBN 978-91-985577-4-9 open access

First published 2021

An electronic version of this book is also available under a Creative Commons (CC-BY-NC-ND) licence, thanks to the support of Lund University, which permits non-commercial use, distribution and reproduction provided the author(s) and Lund University Press are fully cited and no modifications or adaptations are made. Details of the licence can be viewed at https://creativecommons.org/licenses/by-nc-nd/4.0/

The publisher has no responsibility for the persistence or accuracy of URLs for any external or third-party internet websites referred to in this book, and does not guarantee that any content on such websites is, or will remain, accurate or appropriate.

Typeset by
Servis Filmsetting Ltd, Stockport, Cheshire

Contents

List of abbreviations vi

1 Introduction 1

2 The big breakthrough of environmental issues in Sweden, autumn 1967 25

3 The route to the breakthrough, 1948–1967 54

4 How the journalist and the historian came to the environmental issues, 1964–1969 78

5 The environment and the Swedish public, 1967–1968 114

6 The emergence of the modern environmental movements, 1959–1972 144

7 Conflicts and media storms, 1971–1972 168

8 A new history of knowledge 204

Bibliography 218

Index 235

List of abbreviations

ABF	Workers' Education Association (Sweden)
CUF	Youth League of the Centre [Party]
EEC	European Economic Community
FOA	Swedish National Defence Research Institute
IYF	International Youth Federation for the Study and Conservation of Nature
JUF	Jordbrukare Ungdomens Förbund [Agricultural Youth Association]
MIGRI	Miljögruppernas riksförbund [National Association of Environmental Groups]
SCA	Svensk Cellulosa AB
SEK	Swedish kronor
SFU	Sveriges Fältbiologiska Ungdomsförening [Nature and Youth Sweden]
SLU	Swedish University of Agricultural Sciences
SNF	Swedish Society for Nature Conservation
STU	National Board for Technical Development
SUL	National Council of Swedish Youth
UNCTAD	United Nations Conference on Trade and Development

Newspapers and magazines

AB	*Aftonbladet*
Arbt	*Arbetet*
BoT	*Borlänge Tidning*
BT	*Borås Tidning*
DN	*Dagens Nyheter*
Exp	*Expressen*
FB	*Fältbiologen*
GD	*Gefle Dagblad*

List of abbreviations

GHT	Göteborgs Handels- och Sjöfartstidning
GP	Göteborgsposten
GT	Göteborgs-Tidningen
HD	Helsingborgs Dagblad
HT	Hudiksvallstidningen
JP	Jönköpingsposten
KT	Karlstads-Tidningen
KvP	Kvällsposten
LT	Ljungbytidningen
NK	Norrbottenkuriren
NS	Norra Skåne
NT-ÖD	Norrköpings Tidningar-Östergötlands Dagblad
SDS	Sydsvenska Dagbladet
Sf	Smålands folkblad
SkD	Skånska Dagbladet
SLT	Skaraborgs Läns Tidning
SvD	Svenska Dagbladet
UNT	Upsala Nya Tidning
VA	Veckans Affärer
Vf	Värmlands folkblad
VJ	Veckojournalen
Vlt	Vestmanlands läns tidning

1
Introduction

The first United Nations Conference on the Human Environment opened on 6 June 1972. The meeting was held in Stockholm and brought together politicians, researchers, and environmental activists from all over the world. The discussions continued for two weeks and were based on a growing realization that humanity was threatened. Humans themselves were on the verge of destroying their own living environment. Prior to the conference, the report *Only One Earth: The Care and Maintenance of a Small Planet* (1972)[1] was compiled. The front cover featured a picture of the Earth from space. It was a place of life surrounded by a pitch-black void. Humanity's future was at stake.

To the inhabitants of Sweden in 1972, the threat to the planet and to humanity was nothing new. Anyone who regularly read newspapers, listened to the radio, or watched the television news would have encountered the global environmental crisis. School pupils had participated in educational days and watched documentary films. A myriad of small environmental organizations had been founded throughout the country. Intensive debates were being held within and between the five parties in the Swedish parliament, the Riksdag. The European Year of Nature Conservation had been celebrated in 1970, and a year later the concept of the 'Green Movement' (literally 'the Green Wave', *gröna vågen* in Swedish) had been coined. So knowledge about an environmental crisis was definitely circulating in Swedish society in the early 1970s.

Five years earlier, in the summer of 1967, things were different. At that time it was not self-evident that humans were in the process of destroying their own living environment. Of course, many people had noticed the debate about biocides in agriculture and had heard

---

1 Barbara Ward and René Dubos, *Only One Earth: The Care and Maintenance of a Small Planet* (London: Deutsch, 1972).

about the dangers of mercury. Some had even read Rachel Carson's *Silent Spring* (1962). However, in the mid-1960s hardly anyone, even among scientists and politicians, thought in terms of humanity standing on the brink of a global environmental crisis. The various environmental hazards were mainly viewed as separate problems. Each field had its own experts, laws, and technologies. The global systems thinking that characterized the Stockholm Conference in 1972 was not generally prevalent in the summer of 1967. In the next five years a major change occurred in Sweden, as also happened in large parts of the world.[2]

This book explores the major social breakthrough of environmental issues in Sweden. What was it that opened people's eyes to the environmental crisis? When did it happen? Who set the ball rolling? What was done to make it happen? Indeed, what actually happens when knowledge of a kind that has only engaged small groups of people for a long time begins to be noticed in the lives of the vast majority? What happens to society? And what happens to the knowledge?

These issues and themes intersect with our own time. What happened in the years around 1970 was not an isolated chain of events. It had repercussions. To contemporary readers, some of the statements and images I write about will feel strangely familiar.

---

2 John McCormick, *Reclaiming Paradise: The Global Environmental Movement* (London: Belhaven Press, 1989); Ramachandra Guha, *Environmentalism: A Global History* (New York: Longman, 2000); Michael Bess, *The Light-Green Society: Ecology and Technological Modernity in France, 1960–2000* (Chicago, IL: University of Chicago Press, 2003); Kai F. Hünemörder, *Die Frühgeschichte der globalen Umweltkrise und die Formierung der deutschen Umweltpolitik (1950–1973)* (Stuttgart: Steiner, 2004); Jens Ivo Engels, *Naturpolitik in der Bundesrepublik: Ideenwelt und politische Verhaltensstile in Naturschutz und Umweltbewegung 1950–1980* (Paderborn: Schöningh, 2006); Michael Egan, *Barry Commoner and the Science of Survival: The Remaking of American Environmentalism* (Cambridge, MA: MIT Press, 2007); Frank Uekötter, *The Greenest Nation?: A New History of German Environmentalism* (Cambridge, MA: MIT Press, 2014); Adam Rome, *The Genius of Earth Day: How a 1970 Teach-in Unexpectedly Made the First Green Generation* (New York: Hill and Wang, 2013); Joachim Radkau, *The Age of Ecology* (Cambridge: Polity Press, 2014); Sabine Höhler, *Spaceship Earth in the Environmental Age, 1960–1990* (London: Pickering & Chatto, 2015); Perrin Selcer, *The Postwar Origins of the Global Environment: How the United Nations Built Spaceship Earth* (New York: Columbia University Press, 2018); Paul Warde, Libby Robin, and Sverker Sörlin, *The Environment: A History of the Idea* (Baltimore, MD: Johns Hopkins University Press, 2018).

Many of them could have been taken from our own time. The similarities between now and then call for reflection and contemplation. Above all, though, this book aims to provide new historical insights – new because the breakthrough of environmental issues in Sweden in the late 1960s is not a particularly well-known historical process, especially not outside the nation's borders.

What happened in Swedish society was, however, remarkable even from a global perspective. First, the breakthrough of environmental issues in Sweden occurred strikingly early. As early as the autumn of 1967, about a dozen Swedish scientists publicly warned of a global environmental crisis. The researchers had strong ties to the government, the armed forces, and influential media. Unique conditions for moving from knowledge to action existed in Sweden. Second, there was a direct link between the breakthrough of environmental issues in Sweden in the autumn of 1967 and the Stockholm Conference of 1972. The first steps towards the latter were taken on 13 December 1967, when the Swedish UN delegation proposed that a major international environmental conference should be held in the early 1970s. The diplomats acted independently, knowing that people back home were deeply committed to environmental issues.[3]

It may seem surprising that the breakthrough of environmental issues in Sweden has not previously been studied in depth. Given the topicality of environmental and climate issues today, surely masses of historians and social scientists should have traced their roots? Perhaps, though, it is precisely because environmental awareness is so self-evident today that we do not consider that it also has a history. In addition, environmental involvement tends to focus on the present and the future. For people who want to contribute to sustainable social development, the past is not an obvious starting point, especially not the recent past. As a rule, geological and evolutionary time spans overshadow the postwar period.

Nor is the breakthrough of environmental issues a particularly well-known process among people interested in modern history. This is not because the late 1960s have been forgotten by researchers.

---

3 Lars-Göran Engfeldt, *From Stockholm to Johannesburg and Beyond: The Evolution of the International System for Sustainable Development Governance and its Implications* (Stockholm: Government Offices of Sweden, Ministry of Foreign Affairs, 2009), p. 32; Erik Paglia, 'The Swedish Initiative and the 1972 Stockholm Conference: The Decisive Role of Science Diplomacy in the Emergence of Global Environmental Governance', *Humanities and Social Sciences Communications* 8.2 (2021), 1–10.

On the contrary, few postwar periods have been studied so intensely. This interest, however, has primarily been focused on left-wing radicalization and the legendary year 1968. In that time and place, environmental issues were not the centre of attention. It was not until the 1970s, with the Stockholm Conference, the Club of Rome's report on the limits to growth, and the organized resistance to nuclear power, that environmental issues became trendsetting. Or was it? True, from the student left's perspective, that was perhaps the case. But Swedish society was significantly larger than the student left.[4]

A similar argument can be made about the environmental crisis. It was not only of concern to people who were involved in new social movements. In fact, the big breakthrough in Sweden happened before any environmental or alternative movement existed at all. In this book I will argue that the decisive turning point occurred in the autumn of 1967. The driving actors in this process were all part of the social establishment. They were researchers, politicians, editors, and journalists at the major daily newspapers. They had powerful positions and institutional resources. The more small-scale grassroots activism only came later. In the neighbouring countries of Denmark, Norway, and Finland, the breakthrough did not happen until a couple of years later and therefore evolved a different social and political dynamic.[5]

---

4 Kim Salomon, *Rebeller i takt med tiden: FNL-rörelsen och 60-talets politiska ritualer* (Stockholm: Rabén Prisma, 1996); Kjell Östberg, *1968 – när allting var i rörelse: Sextiotalsradikaliseringen och de sociala rörelserna* (Stockholm: Prisma, 2002); Kjell Östberg and Jenny Andersson, *Sveriges historia: 1965–2012* (Stockholm: Norstedts, 2013); Thomas Ekman Jørgensen, *Transformation and Crises: The Left and the Nation in Denmark and Sweden, 1956–1980* (New York: Berghahn, 2008); Henrik Berggren, *68* (Stockholm: Max Ström, 2018).

5 Andrew Jamison, Ron Eyerman, and Jacqueline Cramer with Jeppe Læssøe, *The Making of the New Environmental Consciousness: A Comparative Study of the Environmental Movements in Sweden, Denmark and the Netherlands* (Edinburgh: Edinburgh University Press, 1990); Andrew Jamison and Erik Baark, 'National Shades of Green: Comparing the Swedish and Danish Styles in Ecological Modernisation', *Environmental Values* 8.2 (1999); Peder Anker, 'Den store økologiske vekkelsen som har hjemsøkt vårt land', in John Peter Collett (ed.), *Universitetet i Oslo: 1811–2011* (Oslo: Unipub, 2011); Bredo Berntsen, *Grønne linjer: Natur- og miljøvernets historie i Norge* (Oslo: Unipub, 2011); Tuomas Räsänen, 'Converging Environmental Knowledge: Re-evaluating the Birth of Modern Environmentalism in Finland', *Environment and History* 18.2 (2012); David Larsson Heidenblad, 'En nordisk blick

Introduction 5

I have chosen to characterize what happened in Sweden in the autumn of 1967 as *a social breakthrough of knowledge*. By this I mean a historical process whereby a form of knowledge starts to become very important to many people. The breakthrough of environmental issues is almost an archetypal example of such a historical process. The knowledge which then began to circulate was not new, neither in form nor in content. Nor was it based on any new scientific findings or insights. From the perspective of the history of ideas and of science, it is instead the late 1940s which is the critical turning point. That was when the understanding about a threatened world and humanity was carved out among small – but influential and well-resourced – elite circles at the global level.[6] For most people, this development was out of sight and irrelevant. It would take almost two decades before the looming environmental crisis became part of the lives of the vast majority. For this reason I have chosen to begin my investigation in the autumn of 1967.

The history and circulation of knowledge

My study of the breakthrough of environmental issues seizes on and seeks to develop the new research field concerned with the history of knowledge. This field has emerged during the 2000s and brings together researchers from various historical specialities. In the early 2000s, the discussions were mainly conducted in German-speaking Europe around the concept of *Wissensgeschichte*.[7] Around 2015, however, international interest began to grow, not

---

på det moderna miljömedvetandets genombrott', in Erik Bodensten, Kajsa Brilkman, David Larsson Heidenblad, and Hanne Sanders (eds), *Nordens historiker: En vänbok till Harald Gustafsson* (Lund: Mediatryck, 2018); Peder Anker, *The Power of the Periphery: How Norway Became an Environmental Pioneer for the World* (Cambridge: Cambridge University Press, 2020); Hallvard Notaker, 'Staging Discord: Nordic Corporatism in the European Conservation Year 1970', *Contemporary European History* 29.3 (2020).
6 Warde, Robin, and Sörlin, *The Environment*.
7 Ulrich Johannes Schneider, 'Wissensgeschichte, nicht Wissenschaftsgeschichte', in Axel Honneth and Martin Saar (eds), *Michel Foucault: Zwischenbilanz einer Rezeption* (Frankfurt am Main: Suhrkamp, 2003); Jakob Vogel, 'Von der Wissenschafts- zur Wissensgeschichte der "Wissensgesellschaft"', *Geschichte und Gesellschaft* 30 (2004); Philipp Sarasin, 'Was ist Wissensgeschichte?', *Internationales Archiv für Sozialgeschichte der deutschen Literatur (IASL)* 36 (2011); Daniel Speich Chassé and David Gugerli, 'Wissensgeschichte: Eine Standortbestimmung', *Traverse: Zeitschrift*

least in the Nordic countries. At that time Johan Östling launched the equivalent Swedish term, 'kunskapshistoria', and the following year Peter Burke published *What is the History of Knowledge?* (2016).[8] Since then the international discussion has continued to grow and a number of networks, international journals, and major research projects have been initiated.[9]

However, the rapid growth of the field, as well as its broadly inclusive label, has led to a debate over its value. Critics argue that the history of knowledge is vaguely defined and seems mostly to

---

für Geschichte 19.1 (2012); Jürgen Renn, 'From the History of Science to the History of Knowledge – and Back', *Centaurus: An International Journal of the History of Science & its Cultural Aspects* 57.1 (2015).

8 Johan Östling, 'Vad är kunskapshistoria?', *Historisk tidskrift* 135.1 (2015); Peter Burke, *What is the History of Knowledge?* (Cambridge: Polity Press, 2016).

9 For an overview see Johan Östling, David Larsson Heidenblad, Erling Sandmo, Anna Nilsson Hammar, and Kari H. Nordberg, 'The History of Knowledge and the Circulation of Knowledge: An Introduction', in Johan Östling, Erling Sandmo, David Larsson Heidenblad, Anna Nilsson Hammar, and Kari H. Nordberg (eds), *Circulation of Knowledge: Explorations in the History of Knowledge* (Lund: Nordic Academic Press, 2018); Martin Mulsow and Lorraine Daston, 'History of Knowledge', in Marek Tamm and Peter Burke (eds), *Debating New Approaches to History* (London: Bloomsbury Academic, 2019); Sven Dupré and Geert Somsen, 'The History of Knowledge and the Future of Knowledge Societies', *Berichte zur Wissenschaftsgeschichte* 42.2–3 (2019); Maria Simonsen and Laura Skouvig, 'Videnshistorie: Nye veje i historievidenskaberne', *Temp – Tidskrift for historie* 10.19 (2020); Johan Östling, David Larsson Heidenblad, and Anna Nilsson Hammar, 'Developing the History of Knowledge', in Johan Östling, David Larsson Heidenblad, and Anna Nilsson Hammar (eds), *Forms of Knowledge: Developing the History of Knowledge* (Lund: Nordic Academic Press, 2020); Johan Östling and David Larsson Heidenblad, 'Fulfilling the Promise of the History of Knowledge: Key Approaches for the 2020s', *Journal for the History of Knowledge* 1.1 (2020). Two new book series were launched in 2019: Routledge's 'Knowledge Societies in History' (edited by Sven Dupré and Wijnand Mijnhardt) and Rowman & Littlefield International's 'Global Epistemics' (edited by Inanna Hamati-Ataya). *KNOW: A Journal on the Formation of Knowledge*, whose first issue was published in 2017, has Shadi Bartsch-Zimmer as its editor-in-chief and is the main journal of the Stevanovich Institute on the Formation of Knowledge, University of Chicago. *The Journal for the History of Knowledge* (editors-in-chief: Sven Dupré and Geert Somsen) published its first issue in 2020. In addition, a number of journals devote special issues or theme sections to the history of knowledge, including *Geschichte und Gesellschaft*, *History and Theory*, *History of Humanities*, *Slagmark*, and *Tidskrift for Kulturstudier*.

be a new name for something that researchers have long been working on. These objections have been raised in particular by historians of ideas and science.[10] My own position is that it is a little too early to rule on what original and viable contributions the field does or does not make. Historical scholarship is a slow activity, and it takes a number of years before intellectual ambitions are manifested in pioneering research. I do, however, believe that historians of knowledge should take the objections to the field very seriously. During the 2020s, representatives of the field need to be able to show how their approaches differ from those of others. Their answers both can and should be nuanced and numerous. The history of knowledge is not a theoretical school of thought but rather an integrative field. It is broad enough to accommodate various lines of inquiry and conflicting voices.[11]

My own understanding of the history of knowledge and its potential stems from the perspective of cultural and social history. For me, the focus lies on the social relevance and scope of knowledge. This form of the history of knowledge centres on things that many people have perceived as knowledge, treated as knowledge, and based their actions on as knowledge. By studying this type of phenomenon, I want to contribute to the writing of a broader social history in which knowledge is as self-evident a starting point as politics, economics, or gender. That is not the case today. The lack of interest in knowledge contributes to the fact that even historical processes with far-reaching consequences and obvious contemporary relevance – such as the breakthrough of environmental issues – tend to be marginalized in broader forms of historical writing. The history of knowledge is needed in order to change this oversight.

A key concept in the history of knowledge is circulation. With this, the focus is on how knowledge is shaped and reshaped when it is in motion. The theoretical starting point is that knowledge does

---

10 Lorraine Daston, 'The History of Science and the History of Knowledge', *KNOW: A Journal on the Formation of Knowledge* 1.1 (2017); Suzanne Marchand, 'How Much Knowledge is Worth Knowing? An American Intellectual Historian's Thoughts on the *Geschichte des Wissens*', *Berichte zur Wissenschaftsgeschichte* 42.2–3 (2019); Staffan Bergwik and Linn Holmberg, 'Standing on Whose Shoulders? A Critical Comment on the History of Knowledge', in Östling, Larsson Heidenblad, and Nilsson Hammar (eds), *Forms of Knowledge*.
11 See Östling, Larsson Heidenblad, and Nilsson Hammar (eds), *Forms of Knowledge*.

not exist in any 'pure' form. Knowledge requires channels and bearers in order to move and operate.[12] Johan Östling, with whom I have worked closely to develop a Nordic-based history of knowledge, explains this by saying that knowledge is always formatted.[13] That said, knowledge exists in a constant state of potential change, and mapping and analysing knowledge in motion hence becomes a central research task.[14]

The concept of circulation is currently used in many different ways by researchers both within and outside the history of knowledge field. Its power seems to lie in its ability to offer a concrete alternative to linear dissemination models, which many people reject on theoretical grounds. The concept of circulation thereby complicates questions about how knowledge is produced and becomes important.[15] The most radical voices even question the principle of dissemination and the existence of some kind of starting point for knowledge. From this perspective, production and circulation are inseparable.[16]

I myself, as should be clear by now, am primarily interested in how knowledge moves on a social level. That is, how something

---

12 Philipp Sarasin and Andres Kilcher, 'Editorial', *Nach Feierabend: Zürcher Jahrbuch für Wissensgeschichte* 7 (2011), 8.
13 Östling, 'Vad är kunskapshistoria?', 112.
14 James Secord, 'Knowledge in Transit', *Isis* 95.4 (2004); Anders Ekström, 'Vetenskaperna, medierna, publikerna', in Anders Ekström (ed.), *Den mediala vetenskapen* (Nora: Nya Doxa, 2004); Andreas Daum, 'Varieties of Popular Science and the Transformation of Public Knowledge', *Isis* 100.2 (2009); Solveig Jülich, 'Lennart Nilsson's *A Child is Born*: The Many Lives of a Best-selling Pregnancy Advice Book', *Culture Unbound: Journal of Current Cultural Research* 7.4 (2015).
15 Claude Markovits, Jacques Pouchepadass, and Sanjay Subrahmanyam (eds), *Society and Circulation: Mobile People and Itinerant Cultures in South Asia, 1750–1950* (London: Anthem, 2006); Kapil Raj, *Relocating Modern Science: Circulation and the Construction of Knowledge in South Asia and Europe, 1650–1900* (Basingstoke: Palgrave MacMillan, 2007); Lissa Roberts (ed.), *Local Encounters and Global Circulation*, special issue of *Itinerario* 33.1 (2009); Mary Terrall and Kapil Raj (eds), *Circulation and Locality in Early Modern Science*, special issue of *British Journal for the History of Science* 43.4 (2010); Bernard Lightman, Gordon McOuat and Larry Stewart (eds), *The Circulation of Knowledge Between Britain, India, and China: The Early-Modern World to the Twentieth Century* (Leiden: Brill, 2013); Johan Östling, Erling Sandmo, David Larsson Heidenblad, Anna Nilsson Hammar, and Kari H. Nordberg (eds), *Circulation of Knowledge*.
16 Sarasin and Kilcher, 'Editorial'.

goes from being a matter of concern to some people to becoming of concern to many. Together with Johan Östling, I have chosen to label this understanding of the concept of circulation 'the social circulation of knowledge'. We argue that this more precise definition paves the way for historical studies of key social phenomena which have received far too little attention by scholars. We believe that this concept can contribute to a shift in the centre of gravity – away from a focus on the production and origin of knowledge and towards studies of its circulation and effects.[17]

This view of the history of knowledge does not constitute a radical break with established research traditions. There is great interest in studying publics, media, and public actors, not least within current sociologically inspired research into the history of science.[18] Despite this, comprehensive studies of social breakthroughs of knowledge are unusual, especially in the subject of history, the discipline in which I myself operate and was trained. It is far more common for historians to study discourses and contexts. Textual interpretations and links to the history of ideas are widespread. Whether or not anyone contemporary with the analysed texts even read and cared about them seems to play less of a role. They are perceived as being interesting in and of themselves. It is, of course, possible to believe this. There are good arguments for studying the unnoticed or the unusual. The risk, though, is that an overly strong focus on socially marginal phenomena will cause broader social

---

17 Johan Östling and David Larsson Heidenblad, 'Cirkulation – ett kunskapshistoriskt nyckelbegrepp', *Historisk tidskrift* 137.2 (2017), 279–284; Östling and Larsson Heidenblad, 'Fulfilling the Promise'.
18 Anders Ekström (ed.), *Den mediala vetenskapen* (Nora: Nya Doxa, 2004); Peter Broks, *Understanding Popular Science* (Maidenhead: Open University Press, 2006); Johan Kärnfelt, *Allt mellan himmel och jord: Om Knut Lundmark, astronomin och den publika kunskapsbildningen* (Lund: Nordic Academic Press, 2009); Jonathan Topham, 'Rethinking the History of Science Popularization/Popular Science', in Faidra Papanelopolou, Agustí Nieto-Galan, and Enrique Perdiguero (eds), *Popularizing Science and Technology in the European Periphery 1800–2000* (Farnham: Ashgate, 2009); Agustí Nieto-Galan, *Science in the Public Sphere: A History of Lay Knowledge and Expertise* (Abingdon: Routledge, 2016); Johan Kärnfelt, Karl Grandin, and Solveig Jülich (eds), *Kunskap i rörelse: Kungl. Vetenskapsakademien och skapandet av det moderna samhället* (Gothenburg: Makadam, 2018), pp. 377–438; Solveig Jülich, 'Fosterexperimentets produktiva hemlighet: Medicinsk forskning och vita lögner i 1960- och 1970-talets Sverige', *Lychnos* (2018).

processes to be obscured. I believe that historians should have the ambition to say something about these processes. For this reason we need to devote more care to what we choose to study in detail. The lives and realities of the vast majority deserve more attention.

As is apparent, my understanding of the history of knowledge and circulation is closely intertwined with my interest in the breakthrough of environmental issues. In fact, my theoretical understanding and my empirical research have shaped each other. I began investigating the breakthrough of environmental issues at the same time as I became involved in the history-of-knowledge field, in the autumn of 2014. Since then, my empirical studies and the more general theoretical and methodological discussions have cross-fertilized one another.

Fundamentally, however, I am an empirically orientated and question-driven researcher who wants to find out new things about various developments in the past. The theoretical and methodological approaches which inspire me are the ones that help me move from curiosity to research. For me, both the formation of the history-of-knowledge field and the concept of circulation have fulfilled such functions. Above all, the discussions have driven me to become more concrete and to place more emphasis on analysing actors, networks, types of media, and chronological sequencing. I would go so far as to say that the study of the social circulation of knowledge requires historical research of a relatively high resolution. At an overly aggregated level, the phenomena that I have found to be most important for social circulation processes are not visible: human actions, interactions, media conditions, and historical processes.

The history of knowledge – a methodological intervention

My own move into the history of knowledge has brought me closer to a number of fields and research traditions with which generalist historians are not usually in close contact. These include the sociological history of science, intellectual history, the sociology of knowledge, science and technology studies, and the history of the media, books, and education. In Sweden, a number of these fields are gathered under an umbrella discipline called 'the history of science and ideas'. This is an internationally unusual subject framework which, in addition to Sweden, only exists at a few Norwegian and Danish universities. The discipline partly overlaps with the

Introduction

history of science and intellectual history, but its nature is broader and more eclectic.[19]

Researchers in the above-mentioned fields have long been interested in knowledge. Scientific worlds and practitioners have been a central focus, but they have in no way been studied in isolation from their surrounding society. On the contrary, ever since Ludwik Fleck's exploration of scientific 'thought collectives' and 'thought styles' in the 1930s, researchers have sought in various ways to shed light on the close relationships between science, politics, economics, technology, the media, and social movements.[20] With Sheila Jasanoff's concept of 'co-production', it has been emphasized that knowledge

---

19 Nils Andersson and Henrik Björck (eds), *Idéhistoria i tiden: Perspektiv på ämnets identitet under sjuttiofem år* (Stockholm: Symposion, 2008); Ellen Krefting, Espen Schaanning, and Reidar Asgaard (eds), *Grep om fortiden: Perspektiver och metoder i idéhistorie* (Oslo: Cappelen Damm Akademisk, 2017); Mikkel Thorup, *Hvad er idéhistorie?* (Aarhus: Slagmark forlag, 2019); Anton Jansson, 'Things are Different Elsewhere: An Intellectual History of Intellectual History in Sweden', *Global Intellectual History* 6.1 (2021); Anton Jansson and Maria Simonsen, 'Kunskapshistoria, idéhistoria och annan historia: En översikt i skandinaviskt perspektiv', *Slagmark* 81 (2020).

20 Cf.: Ludwik Fleck, *Entstehung und Entwicklung einer wissenschaftlichen Tatsache: Einführung in die Lehre vom Denkstil und Denkkollektiv* (Basel: Schwabe, 1935); Thomas S. Kuhn, *The Structure of Scientific Revolutions* (Chicago, IL: University of Chicago Press, 1962); Donna Harraway, 'Situated Knowledges: The Science Question in Feminism and the Privilege of Partial Perspective', *Feminist Studies* 14.3 (1988); Robert K. Merton, *On Social Structure and Science* (Chicago, IL: University of Chicago Press, 1996); Sheila Jasanoff, Gerald E. Markle, James C. Peterson, and Trevor Pinch (eds), *Handbook of Science and Technology Studies* (Thousand Oaks, CA: SAGE, 1995); Sven Widmalm (ed.), *Vetenskapsbärarna: Naturvetenskapen i det svenska samhället 1880–1950* (Hedemora: Gidlunds, 1999); Michel Foucault, *The Essential Foucault: Selections from Essential Works of Foucault, 1954–1984* (New York: New Press, 2003); Jan Golinski, *Making Natural Knowledge: Constructivism and the History of Science* (Chicago, IL: University of Chicago Press, 2005); Robert Fox, 'Fashioning the Discipline: History of Science in the European Intellectual Tradition', *Minerva* 44.4 (2006); Harry Collins and Robert Evans, *Rethinking Expertise* (Chicago, IL: University of Chicago Press, 2007); Sven Widmalm (ed.), *Vetenskapens sociala strukturer: Sju historiska fallstudier om konflikt, samverkan och makt* (Lund: Nordic Academic Press, 2008); Jon Agar, *Science in the Twentieth Century and Beyond* (Cambridge: Polity, 2012); Lorraine Daston, 'Science, history of', in James D. Wright (ed.) *International Encyclopedia of the Social and Behavioral Sciences* (Oxford: Elsevier, 2015); Lynn K. Nyhart, 'Historiography of the History of Science', in Bernard Lightman (ed.), *A Companion to the History*

and social development exist in an almost symbiotic relationship with each other. This has paved the way for critical analyses of the power relationships in which all knowledge production and circulation are involved.[21] Questions about how scientific legitimacy is created and maintained have also attracted great interest. Drawing on Thomas Gieryn's scholarship, special emphasis has been placed on the various forms of 'boundary work' that scientific actors employ in order to assert authority and gain influence. This work may entail marking territories and maintaining dividing lines, but also exceeding boundaries and emphasizing the relevance of knowledge in new fields, including political ones.[22] A third important concept is 'network' or 'networking', which in scientific studies includes both people and various material objects. In this case, the focus is not on the boundaries but on the relationships between various actors and things. Researchers such as Bruno Latour analyse how networks enable and legitimize knowledge and can make it move between different contexts.[23]

---

of Science (Chichester: John Wiley & Sons, 2016); James Poskett, 'Science in History', *Historical Journal* 63.2 (2020).

21 Sheila Jasanoff, 'Ordering Knowledge, Ordering Society', in Sheila Jasanoff (ed.), *States of Knowledge: The Co-production of Science and Social Order* (London: Routledge, 2004); Staffan Bergwik, Michael Godhe, Anders Houltz, and Magnus Rodell (eds), *Svensk snillrikhet? Nationella föreställningar om entreprenörer och teknisk begåvning 1800–2000* (Lund: Nordic Academic Press, 2014); Anna Tunlid and Sven Widmalm (eds), *Det forskningspolitiska laboratoriet: Förväntningar på vetenskapen 1900–2010* (Lund: Nordic Academic Press, 2016).

22 Thomas Gieryn, *Cultural Boundaries of Science: Credibility on the Line* (Chicago, IL: University of Chicago Press, 1999); Anna Tunlid, *Ärftlighetens gränser: Individer och institutioner i framväxten av den svenska genetiken* (Lund: Department of Cultural Sciences, History of Ideas Unit, 2004); Staffan Wennerholm, *Framtidsskaparna: Vetenskapens ungdomskultur vid svenska läroverk 1930–1970* (Lund: Arkiv, 2005); Per Wisselgren, 'Vetenskap och/ eller politik? Om gränsteorier och utredningsväsendets vetenskapshistoria', in Bosse Sundin and Maria Göransdotter, *Mångsysslare och gränsöverskridare: 13 uppsatser i idéhistoria* (Umeå: Umeå University, 2008).

23 Bruno Latour, *Science in Action: How to Follow Scientists and Engineers through Society* (Cambridge, MA: Harvard University Press, 1987); John Law and John Hassard, *Actor Network Theory and After* (Oxford: Blackwell, 1999); Nina Wormbs, *Vem älskade Tele-X? Konflikter om satelliter i Norden 1974–1989* (Hedemora: Gidlunds, 2003); Sven Widmalm and Hjalmar Fors (eds), *Artefakter: Industrin, vetenskapen och de tekniska nätverken* (Hedemora: Gidlunds, 2004); Bruno Latour, *Reassembling the Social: An Introduction to Actor-Network-Theory* (Oxford: Oxford University Press, 2005).

Against this background, it is perhaps not surprising that the rapid emergence of the history-of-knowledge field has been regarded with misgivings in some camps. What exactly is new? How do history-of-knowledge perspectives relate to established discussions and theory constructions? Upon whose shoulders do historians of knowledge actually stand? Indeed, what new insights can the field contribute?[24] As Suzanne Marchand expressed it, might it not just be a matter of 'old wine in slightly stretched wine skins'? Is a new name really needed for something that so many researchers have already devoted so much time to?[25]

I want to argue here that the history of knowledge – in the version that Johan Östling and I have tried to develop – does in fact contribute a new orientation. It involves a methodological intervention which aims to generate new questions and lines of research and, by extension, new insights into various key social processes, for example the emergence of modern environmental awareness. Our focus lies on questions about what happens when various forms of knowledge become matters of social concern and intervene in many people's lives. What makes knowledge circulate through society? How does it happen, and what are the consequences? These questions are in themselves not new. However, they neither are nor have been of primary consideration in the fields discussed above.[26]

Paradoxically, the methodological intervention is a consequence of the fact that the foremost area of interest for the new history of knowledge is not knowledge and its epistemic conditions but rather the broader development of society, of which various forms of knowledge are one important aspect. To quote Simone Lässig, the overall goal of those who examine knowledge in circulation is to gain a 'better understanding of societies'.[27] This ambition is rooted in a fundamental perspective based on social and cultural history, plus a programmatic interest in processes that intervene in the lives of the vast majority. The aim is to contribute more wide-ranging histories of society which include many different voices and

---

24 Staffan Bergwik, 'Kunskapshistoria: Nya insikter?', *Scandia* 84.2 (2018); Bergwik and Holmberg, 'Standing on Whose Shoulders?'
25 Marchand, 'How Much Knowledge'.
26 Secord, 'Knowledge in Transit'.
27 Simone Lässig, 'The History of Knowledge and the Expansion of the Historical Research Agenda', *Bulletin of the German Historical Institute* 59 (2016), 43.

perspectives. The history of knowledge is thus an attempt to move research in new directions.

In order to achieve this methodological intervention, the study of 'the social circulation of knowledge' is key. This focus directs our attention at when, how, why, and with what consequences knowledge is updated and makes an impression on many people's lives. Public spheres, media forums, influential organizations, and leading actors obviously play a central role; but it is also important to study other actors, groups, and audiences. How can we study the social circulation of knowledge if we only look at those actors who are in the spotlight? In order to study knowledge empirically as a far-reaching social phenomenon, we must look more widely and not become stuck in close-up studies of the most obvious historical actors, organizations, and arenas.

That said, it is important not to neglect influential elites, networks, and institutions. Their activities and significance need to be empirically investigated and assessed. The same applies to questions about how knowledge is set in motion and what happens to it when it starts to circulate widely. In this regard, it must be emphasized that we cannot assume in advance that knowledge is never 'spread' or 'seeps down' from elites to the majority without undergoing fundamental change. This may well be the case, but the circulation concept as employed by the history of knowledge makes it an open empirical question.

This pragmatic approach also applies to my view of the question of what knowledge is, and what knowledge it is that I am actually studying. Here I agree with Jürgen Renn's argument that studies of the history of knowledge should try to find a middle ground between knowledge as pertaining to a category of historical actors and knowledge as a purely analytical category. Both of these extremes cause problems. The former may lead to a radically subjectivist and relativistic position which makes it impossible to compare phenomena across time and space. The latter could become anachronistic, simplistic, and empirically difficult to use in historical studies. Renn's way out of this dilemma is to regard studies of the history of knowledge as explorations of both the past and 'the nature of knowledge itself'.[28] In order for the potential of the history-of-knowledge field to be fulfilled, however, this search must be clarified and discussed,

---

28 Jürgen Renn, *The Evolution of Knowledge: Rethinking Science for the Anthropocene* (Princeton, NJ: Princeton University Press, 2020), p. 11.

Introduction 15

so that a better collective understanding of what knowledge is and has been can be developed. In this respect, as Lorraine Daston and other critics have pointed out, historians of knowledge have things to learn from historians of science.[29]

My own practical entry point into this study has been to investigate publicly expressed claims of knowledge about a looming environmental crisis that have had widespread impact, for example via bestselling books. This has been one way of accessing the social circulation of knowledge. I have focused on mapping and analysing how the books were discussed in their own time rather than on analysing their contents. The future-focused expertise that has been accorded to certain actors has been of particular importance. I have subsequently supplemented this initial empirical starting point by following up various threads, actors, organizations, and relationships. With this approach, it has not been possible to find a straightforward definition of what was perceived and handled as knowledge. At different times and for different actors, the main focus and understanding of knowledge about the environmental crisis looked quite different. This is not to say that no patterns exist. However, these patterns are the result of my examinations and analyses rather than something I knew beforehand. For example, with regard to questions about 'knowledge dissemination' and 'the importance of elite actors', my empirical results have taken me in different directions than I had imagined on the basis of my prior theoretical understanding. I do not regard this as a problem but rather as a sign that the history of knowledge actually does function in a way that can give us new insights. What, then, is the significance of this methodological intervention for research on the breakthrough of environmental issues?

The ecological turn

In international environmental history research, the late 1960s and early 1970s are referred to as 'the ecological turn' or 'the ecological moment'.[30] This was when environmental issues seriously began to make their presence felt in politics, culture, and social life around the world. Characteristic of this development was that many scientists, such as Barry Commoner in the United States, Jean Dorst in France,

---

29 Daston, 'The History of Science'; Bergwik and Holmberg, 'Standing on Whose Shoulders?'
30 Jens Ivo Engels, 'Modern Environmentalism', in Frank Uekötter (ed.), *The Turning Points of Environmental History* (Pittsburgh, PA: Pittsburgh University

and Hans Palmstierna in Sweden, began to regard it as their task to intervene directly in the social debate in order to try to steer political development down new paths. It was also at this time that the modern environmental movements began to emerge. International organizations such as Friends of the Earth (1969) and Greenpeace (1971) saw the light of day, at the same time as older nature-conservation organizations such as the Sierra Club (1892) and the Swedish Society for Nature Conservation (1909) began to orientate themselves in new directions. In many countries, authorities with special responsibility for environmental issues were established and the legislation in this field was expanded and strengthened. Steps also began to be taken towards deeper international cooperation and agreements.[31]

When the UN's first environmental conference was held in Stockholm in 1972, the theme was 'one world'. It was a vision that ran counter to the way in which the world was generally perceived and functioned at that time. The Cold War was still going on between East and West, and countries in the so-called Third World were recurring arenas for ideological and military confrontations between the blocs. The road to the Stockholm Conference was also lined with high-level political complications. The reason was that East Germany was not allowed to participate because it was not a member of the UN. Most of the Eastern bloc therefore boycotted the event. The only communist countries present were Yugoslavia, China, and Romania. At the conference itself, however, the focus ended up being on the North–South conflict. The Western world's efforts to deal with environmental degradation and overpopulation were pitted against the developing countries' desire for industrialization and prosperity.[32] The inaugural speech by Sweden's prime minister, Olof Palme, was also controversial. He highlighted 'the tremendous

---

Press, 2010), pp. 119–120; Holger Nehring, 'Genealogies of the Ecological Moment: Planning, Complexity and the Emergence of "the Environment" as Politics in West Germany, 1949–1982', in Sverker Sörlin and Paul Warde (eds), *Nature's End: History and the Environment* (Basingstoke: Palgrave Macmillan, 2009).

31 McCormick, *Reclaiming Paradise*; Guha, *Environmentalism*; Frank Zelko, *Make it a Green Peace!: The Rise of Countercultural Environmentalism* (New York: Oxford University Press, 2013); Radkau, *The Age of Ecology*; Peter Dauvergine, *Historical Dictionary of Environmentalism*, 2nd edition (London: Rowman & Littlefield, 2016; first published in 2009).

32 McCormick, *Reclaiming Paradise*, pp. 88–105; Engfeldt, *From Stockholm to Johannesburg*; Anne Egelston, *Sustainable Development: A History* (Dordrecht: Springer Netherlands, 2013).

destruction caused by extensive indiscriminate bombing' and 'the large-scale use of bulldozers and herbicides'.[33] Although it was not stated explicitly, there was no doubt that his critical remarks were aimed at US conduct in Vietnam, which at that time was described in terms of an 'ecocide'. Palme's speech was not appreciated in Washington. A spokesperson for the US State Department said that 'deep unease' was felt over the way that the prime minister of the host country had raised this issue, which had nothing to do with the environmental-protection conference.[34] The UN conference was also sharply criticized by the new environmental movements. They argued that the event was a top-down and inadequate symbolic act. As a result, parallel alternative environmental conferences were organized, such as the radical left-wing People's Forum.[35]

In reality, the Stockholm Conference thus highlighted the many and profound contradictions that characterized 'the one world' in 1972. Knowledge of an ongoing environmental crisis was indeed circulating globally at this time, but it was understood and handled in disparate ways within various power blocs and countries. Of course, this had also been the case before the Stockholm Conference. If we look at years like 1970, 1967, or 1963, the differences were at least as great as in 1972. In order to understand and explain the ecological turn, we therefore need studies of how the process developed in various societies with differing conditions, problems, and agendas. This will also make it possible to show what the global influence processes looked like in practice, and to analyse the chain reaction that made environmental issues a global political concern.[36]

From such a perspective, it is apparent that the first and strongest driving forces behind the ecological turn came from the United States. As far back as the late 1940s, scientists such as William Vogt and Fairfield Osborn had already begun to influence the social debate. They warned that overpopulation and looting of the planet's resources could eventually lead to a global civilizational collapse.[37] The warnings were heeded elsewhere in the West. In Sweden, for

---

33 Anon., 'USA-kritik mot Palme: Oöverlagt och ensidigt om Vietnam', *Svenska Dagbladet (SvD)*, 7 June 1972.
34 Anon., 'USA-kritik mot Palme'.
35 Egelston, *Sustainable Development*, p. 69.
36 Radkau, *The Age of Ecology*, p. 79.
37 Thomas Robertson, *The Malthusian Moment: Global Population Growth and the Birth of American Environmentalism* (New Brunswick: Rutgers, 2012), pp. 36–60; Radkau, *The Age of Ecology*, p. 91.

example, they were picked up by food researcher Georg Borgström, who doggedly spread them in Scandinavia.[38] The next important American impetus came in the early 1960s with Rachel Carson's *Silent Spring* (1962). It focused attention on the dangers of chemical pesticides and sparked fierce debates between nature-conservation interests, industry representatives, and government agencies.[39] Carson's book, however, did not give rise to any grassroots movement, at least not right away. Instead, the birth moment of the American environmental movement was the holding of the first Earth Day on 22 April 1970. An estimated 20 million Americans participated in the event which channelled and strengthened the growing environmental involvement, not least among young school pupils and college students. In the words of environmental historian Adam Rome, Earth Day created 'the first green generation'.[40]

From a global perspective, however, the American celebration of Earth Day was less important. For example, Swedish media did not report on it at all. In the UK and Germany interest seems to have been somewhat greater, but environmental historian Frank Uekötter nevertheless states that Earth Day was 'a purely American event'.[41] Timewise, though, the event coincided with a similar initiative under the auspices of the Council of Europe. The Council had designated 1970 as the European Conservation Year, and political attempts were made throughout the continent to raise awareness of

---

38 Björn-Ola Linnér, *The World Household: Georg Borgström and the Postwar Population–Resource Crisis* (Linköping: Tema University, 1998); Sunniva Engh, 'Georg Borgström and the Population–Food Dilemma: Reception and Consequences in Norwegian Public Debate in the 1950s and 1960s', in Johan Östling, Niklas Olsen, and David Larsson Heidenblad (eds), *Histories of Knowledge in Postwar Scandinavia: Actors, Arenas, and Aspirations* (Abingdon: Routledge, 2020).
39 Thomas Dunlap, *DDT: Scientists, Citizens, and Public Policy* (Princeton, NJ: Princeton University Press, 1981); Linda Lear, *Rachel Carson: Witness for Nature* (New York: Holt, 1997); Gary Kroll, 'The "Silent Springs" of Rachel Carson: Mass media and the origins of modern environmentalism', *Public Understanding of Science* 10.4 (2001); David Vail, *Chemical Lands: Pesticides, Aerial Spraying, and Health in North America's Grasslands since 1945* (Tuscaloosa, AL: University of Alabama Press, 2018).
40 Rome, *The Genius of Earth Day*; David Larsson Heidenblad, 'Så uppstod den första "gröna generationen"', *SvD*, 20 April 2020.
41 Thorsten Schulz, *Das 'Europäische Naturschutzjahr 1970': Versuch einer europaweiten Umweltkampagne* (Berlin: Wissenschaftszentrum für Sozialforschung, 2006), pp. 22–23; Uekötter, *The Greenest Nation?*, p. 82.

environmental problems. The results were meagre. The European Conservation Year did not become a catalyst for grassroots involvement in Europe. The environmental movements on this side of the Atlantic emerged in other ways.[42]

The difficulties of coming together around environmental issues at the international level are further illustrated by the Nordic Nature Conservation Day, which was held on 6 September 1970. Sweden, Norway, Denmark, and Finland had collaborated to orchestrate a joint demonstration for the environment. The plan was that hundreds of warning beacon fires would be lit across the Nordic region and culminate in a torchlight procession in Oslo, concluding with fireworks. However, the national committees had quite diverse mandates, compositions, and wishes. For example, the Norwegian one had links to radical forces within the emerging environmental movement, whereas the Swedish one actively distanced itself from them. The Danish and Finnish participation seems to have been lukewarm and characterized by tight budgets.[43]

Nordic Nature Conservation Day thereby reflected the differing paths of development followed by the ecological turn in the Nordic region. In 1970 there existed an environmental policy establishment in Sweden centred around the National Environment Protection Board [Statens Naturvårdsverk, now the Swedish Environmental Protection Agency – translator's note]. The board had been established in June 1967 as the first authority in the world of its kind. Behind the move lay the Social Democratic government, which had ruled Sweden since the 1930s. In Norway and Denmark, the Social Democrats were in opposition at this time. Similar environmental protection authorities would not be established there until 1971 (Denmark) and 1972 (Norway). In Finland it took until 1983. However, it was not only on political and administrative grounds that the countries differed. Even more important was the fact that the Swedish scientific research community was large and resource-rich and became involved in the issues early on. In the wake of the biocide debate, the 1964 government enquiry into natural resources was commissioned to survey the environmental situation in Sweden.

---

42 Uekötter, *The Greenest Nation?*, p. 82; Jan-Henrik Meyer, 'From Nature to Environment: International Organizations and Environmental Protection before Stockholm', in Wolfram Kaiser and Jan-Henrik Meyer (eds), *International Organizations and Environmental Protection* (Oxford: Berghahn, 2017); Notaker, 'Staging Discord'.
43 Notaker, 'Staging Discord'.

Nothing similar happened in the other Nordic countries. The early Swedish warning voices, such as Georg Borgström and Hans Palmstierna, also came to play important roles in the neighbouring countries. There, too, it was not necessarily established scientists who were the most important actors in the social circulation of knowledge. In Norway the philosopher Arne Naess and the advertising executive Erik Dammann came to play a central role during the 1970s. In Denmark it was the student activists within the environmental movement NOAH (1969) who set the tone.[44]

All in all, this meant that the ecological turn acquired a special dynamic in Sweden. In many ways, the similarities were greater with the United States than with Sweden's neighbours and the rest of Western Europe. One important difference, however, was that the Swedish scientists were significantly closer to the centre of national political power than their American counterparts. Barry Commoner and Paul Ehrlich had neither any parliamentary platform nor access to a grassroots movement.[45] What consequences did this have? How did the environmental turn in Sweden happen?

To investigate this, the methodological intervention of the history of knowledge is particularly helpful: this approach enables an examination of the process of change from a wide-ranging social perspective which also allows for a focus on the historical actors. My study, however, does not only examine the most obvious actors, the scientific warning voices and the environmental activists. They are certainly included and important; but they do not stand alone. In order to study the social circulation of knowledge, the net must be cast more widely.

Three knowledge actors

A central point of this book is that historical actors were drivers of and within the breakthrough of environmental issues in Sweden. In

---

[44] Jamison, Eyerman, and Cramer, *The Making of the New Environmental Consciousness*; Jamison and Baark, 'National Shades of Green'; Anker, 'Den store økologiske vekkelsen'; Berntsen, *Grønne linjer*; Räsänen, 'Converging Environmental Knowledge'; Simone Müller, 'Corporate Behaviour and Ecological Disaster: Dow Chemical and the Great Lakes Mercury Crisis, 1970–1972', *Business History* 60.3 (2018); Larsson Heidenblad, 'En nordisk blick'; Anker, *The Power of the Periphery*.

[45] Egan, *Barry Commoner and the Science of Survival*; Robertson, *The Malthusian Moment*, pp. 126–151.

my view, the social breakthrough of knowledge occurred because specific people did specific things at specific times, which triggered chain reactions. I want to make this historical dynamic visible in my presentation. For this reason, I have chosen to highlight three knowledge actors in particular: the chemist Hans Palmstierna, the journalist Barbro Soller, and the historian Birgitta Odén. All three were born in the 1920s and were in the midst of their lives and careers during the years I study. They all contributed to the breakthrough of environmental issues, which in turn led to new directions in their own lives and those of others.

Best known in his day was Hans Palmstierna. He was an associate professor of chemistry, did laboratory work at the Karolinska Institute in Stockholm, and worked at the National Bacteriological Institute. He also wrote regularly for the liberal *Dagens Nyheter*, Sweden's largest and most prestigious broadsheet. Palmstierna came from an old noble family but held strong socialist convictions and was active in the Social Democratic Party. In the autumn of 1967, he published the polemical book *Plundring, svält, förgiftning* [Plundering, famine, poisoning]. It came to have a huge impact. Palmstierna became the first truly major environmental debater in Sweden. In early 1968, he left his academic career to start working at the National Environment Protection Board. There he combined his new job with ambitious popular education efforts and political assignments for the Social Democrats. On his death in 1975, it was said that he was the person who 'awakened our awareness' and 'really got the environmental debate going'.[46]

Hans Palmstierna is not an unknown historical actor, but his personal archive has not been used before.[47] It includes a rich collection of letters, diaries, and press clippings, which allow his actions to be studied in detail. The many letters enable me to demonstrate

---

46 Inger Marie Opperud, 'Hans Palmstierna funnen drunknad', *Expressen (Exp)*, 28 May 1975; Björn Berglund, 'Han väckte vårt medvetande', *Dagens Nyheter* (DN), 29 May 1975; Bo Melander, 'Palmstierna – väckarklocka i flera viktiga miljöfrågor', *Göteborgsposten* (GP), 29 May 1975.

47 Jamison, Eyerman, and Cramer, *The Making of the New Environmental Consciousness*, pp. 20–22; Jonas Anshelm, *Socialdemokraterna och miljöfrågan: En studie av framstegstankens paradoxer* (Stockholm: Brutus Östling's Symposion, 1995), pp. 16–27; Nikolas Glover, 'Unity Exposed: The Scandinavia Pavilions at the World Exhibitions in 1967 and 1970', in Jonas Harvard and Peter Stadius (eds), *Communicating the North: Media Structures and Images in the Making of the Nordic Region* (Burlington, VT: Ashgate, 2013), pp. 232–234.

how knowledge circulated in Swedish society at this time. Palmstierna corresponded not only with scientists, fellow political party members, trade-union representatives, and educational associations, but also with clergy, students, bank managers, journalists, and upper-secondary-school pupils. My study will also show that Palmstierna's position underwent several changes. In the autumn of 1967, he went from being an obscure scientist to the role of a unifying and exalted expert. In the early 1970s, though, he became a controversial person whom many turned against. Here it becomes possible to highlight how environmental knowledge and expertise about the future came to be transformed when the environmental debate was converted into political actions in the 1970s.[48]

Barbro Soller was a different type of knowledge actor from Hans Palmstierna. She was neither a researcher nor a politician but a journalist. Her texts were not published on the culture and editorial pages (the forums where the press mainly conducted the environmental debate); they were more like news reporting. Hired by *Dagens Nyheter* in 1964, she developed into Sweden's first environmental journalist in its pages. Her big breakthrough came with the reportage series 'Nya Lort-Sverige' [New filth-Sweden] in the spring of 1968. In the series, she travelled around Sweden to document environmental destruction and littering. The following year, the series was brought out as a reportage book by the publisher who was behind Palmstierna's *Plundring, svält, förgiftning*.

At that time, Barbro Soller's investigative environmental journalism was something completely new. She has been the object of scholarly interest before; in particular, her transition to the TV medium in the early 1970s has attracted attention.[49] I will instead study an earlier phase of her journalistic career. This can now be done because the newspaper material has been digitized and is full-text searchable.[50] I am therefore able to survey her activities and analyse her transition from general reporter to investigative environmental journalist. I

---

48 Paul Warde and Sverker Sörlin, 'Expertise for the Future: The Emergence of Environmental Prediction c.1920–1970', in Jenny Andersson and Eglė Rindzevičiūtė (eds), *The Struggle for the Long-term in Transnational Science and Politics* (New York: Routledge, 2015).
49 Monika Djerf Pierre, *Gröna nyheter: Miljöjournalistiken i televisionens nyhetssändningar 1961–1994* (Gothenburg: Department of Journalism, Media and Communication, University of Gothenburg, 1996).
50 For a discussion see: David Larsson Heidenblad, 'The Emergence of Environmental Journalism in 1960s Sweden: Methodological Reflection on

can hence place her within a larger history-of-knowledge context and illuminate that context through her.

Alongside Hans Palmstierna and Barbro Soller, Birgitta Odén may appear to be an atypical example. What does a historian, who became a professor by studying sixteenth-century state finances, have to do with the breakthrough and social circulation of knowledge about environmental issues? Quite a lot, actually. In the spring of 1967, Odén was invited to a meeting at the Swedish National Defence Research Institute (FOA) (now the Swedish Research Agency). The initiator of this meeting was the director-general and head of FOA, Martin Fehrm. He perceived the environmental problems as a security threat and felt that they could not be reduced to a scientific and technical issue. They involved people's actions and political decision-making processes. For that reason, knowledge based on the humanities and social sciences was needed. Odén became the driving force behind the setting up of such research. In parallel with this, she tried to launch an environmental history research group at the history department in Lund, mainly by involving young students. In the summer of 1968, together with political scientists and economists, she submitted a major research application, but it was never granted.

Birgitta Odén's work shows how the breakthrough of environmental issues changed the life of one individual towards the end of the 1960s. It also indicates how a person in her position could inspire others and establish new directions. Some of the students she involved came to devote their lives to these issues. But her story also indicates the importance of networks and personal connections. Her younger brother was the soil chemist Svante Odén, an associate professor at the Swedish University of Agricultural Sciences in Uppsala. It was he who made the scientific discovery of the environmental hazard that was acid rain and who made the Swedish public aware of it through an article in *Dagens Nyheter* in the autumn of 1967. At the inaugural meeting at FOA in May 1967, both siblings attended.[51]

Hans Palmstierna, Barbro Soller, and Birgitta Odén are not the only actors I study. Through them I reach others, such as the committed layman Sören Gunnarsson, the young environmental activist

---

Working with Digitalized Newspapers', in Östling, Olsen, and Larsson Heidenblad (eds), *Histories of Knowledge in Postwar Scandinavia.*

51 David Larsson Heidenblad, 'Miljöhumaniora på 1960-talet? Birgitta Odéns miljöhistoriska initiativ och skissernas historiografi', *Scandia* 85.1 (2019).

Wolter Arnberg, the secondary-school teacher Kerstin Hägg, and the Lund University student Lars J. Lundgren. The reason why I selected Palmstierna, Soller, and Odén is because they are central enough actors – and different enough in their missions – that a picture of Swedish society at that time can emerge through them. As far as I know, they did not know one another particularly well, but they definitely knew of one another. They were connected via people like Svante Odén and forums like *Dagens Nyheter*. There was also a historical simultaneity. In the autumn of 1967, all three were deeply involved in what came to be the breakthrough of environmental issues. The historical process thus becomes visible through them. That said, we will now turn our gaze to the eventful autumn of 1967.

2
The big breakthrough of environmental issues in Sweden, autumn 1967

In the autumn of 1967, the Swedish environmental debate changed. At that time, a number of prominent scientists publicly warned of an impending global catastrophe. The impact was powerful. There was talk of a general awakening. The press, radio, and television reported on mercury-poisoned fish, biocides, and acid rain. In the apt words of Lars J. Lundgren, it was as if a new continent of problems had been discovered. Previously, various environmental hazards had been regarded as individual islands of problems. Now more and more people were beginning to see them as connected.[1]

At the centre of this development was Hans Palmstierna. That October, he published a debate book in paperback format: *Plundring, svält, förgiftning* [Plundering, famine, poisoning]. He wanted it to 'awaken and create clarity' about the human situation. Palmstierna argued that there was an urgent need to act 'before the hourglass expired for humanity'.[2] It was characteristic of Palmstierna that he linked environmental destruction with other global issues, such as world poverty, war, and overpopulation. He emphasized that the Earth was a small sphere with a limited surface area. For that

---

1 Jan Thelander and Lars J. Lundgren, *Nedräkning pågår: Hur upptäcks miljöproblem? Vad händer sedan?* (Solna: National Environment Protection Board, 1989); Martin Bennulf, *Miljöopinionen i Sverige* (Lund: Dialogos, 1994). This chapter is based on David Larsson Heidenblad, 'Mapping a New History of the Ecological Turn: The Circulation of Environmental Knowledge in Sweden 1967', *Environment and History* 24.2 (2018) and David Larsson Heidenblad, 'Överlevnadsdebattörerna: Hans Palmstierna, Karl-Erik Fichtelius och miljöfrågornas genombrott i 1960-talets Sverige', in Fredrik Norén and Emil Stjernholm (eds), *Efterkrigstidens samhällskontakter* (Lund: Mediehistoriskt arkiv/Media History Archives, 2019).

2 Hans Palmstierna, *Plundring, svält, förgiftning* (Stockholm: Rabén & Sjögren, 1967). Back cover text.

reason, we must 'stop the population growth if humanity is to survive'.[3] Palmstierna predicted an apocalyptic time through which humanity must pass 'in order to be healed into common sense and humility in the face of the implacable laws that prevail in all living things'.[4]

Palmstierna's tone of voice was loud and strong but not unique. In the preface to another discussion book published in paperback in the autumn of 1967, *Människans villkor: En bok av vetenskapsmän för politiker* [The predicament of man: a book by scientists for politicians], Karl-Erik Fichtelius, professor of histology at Uppsala University, wrote that '[d]oomsday prophets have existed for as long as there have been humans. What is new is that now every politically aware scientist can come forward as a doomsday prophet.'[5]

Fichtelius was the editor and initiator of *Människans villkor*. The book assembled twelve major researchers, including the physicist Hannes Alfvén, the economist Gunnar Myrdal, and the food researcher Georg Borgström. Published in December, the book caused an intense debate about the relationship between science and politics. However, one of the participants, Swedish Academy member Lars Gyllensten, had already given a high-profile radio lecture in October. It emphasized that the basic biological conditions for human existence were in the process of being destroyed. Gyllensten said that knowledge about this serious situation was widespread, but that it was not being taken seriously. It would require a 'conscious, effective and unsentimental retraining of us all' if the global problems were to be solved.[6]

Humanity's survival was central to the Swedish environmental debate in the autumn of 1967. At the same time, however, environmental issues were also being discussed in a lower key. The Social Democrat Valfrid Paulsson held a crucial position in this context. In July that year, he had been appointed director-general of the National Environment Protection Board. Most scientific researchers also adopted this more low-key approach. That was noticeable not least in the report by the 1964 government enquiry into natural

---

3 Palmstierna, *Plundring, svält, förgiftning*, p. 15.
4 *Ibid.*, p. 9.
5 Karl-Erik Fichtelius, 'Preface' in Karl-Erik Fichtelius (ed.), *Människans villkor: En bok av vetenskapsmän för politiker* (Stockholm: Wahlström & Widstrand, 1967), p. 5.
6 'Angeläget', Sveriges Radio, 21 October 1967; Lars Gyllensten, 'Politik och undanflykt', *Kvällsposten* (*KvP*), 3 November 1967.

resources, which was submitted in November 1967. Behind the two-volume report was the country's scientific expertise in the environmental field. The enquiry had surveyed the state of knowledge about, and the extent of, various forms of poisoning and pollution. The perspective was national rather than global. Better planning and more research resources were requested. There was no talk of having to retrain people or of establishing a global government.[7]

Even so, the report did point out that environmental problems were not a strictly national matter. This was especially true of Svante Odén's discovery of acid rain. The emissions occurred on the main European continent, but the rain fell on Sweden. International cooperation was necessary to deal with the problem. It is noteworthy how the realization that there was danger afoot was made public: it was presented in an article on *Dagens Nyheter's* culture page, written by Svante Odén himself. The article was part of the broadsheet's series 'Miljö för framtiden' [Environment for the future], which ran from September to December. The series afforded leading scientists space to present and discuss various environmental problems in depth. Odén's article immediately put acidification on to the day-to-day political agenda.[8]

The national side of the environmental debate also included mercury poisoning. In the summer of 1967, it had been discovered that the fish in many Swedish lakes contained high levels of mercury. A ban on selling the fish was introduced. For commercial fishermen in Lake Vänern, Sweden's largest lake, the ban meant unemployment. The events attracted a lot of media attention. They showed that environmental toxins were a direct threat to human lives and livelihoods.

The eventful autumn of 1967 hence featured many themes, directions, and voices. Because of this, Hans Palmstierna came to play a special role. He spoke about humanity's survival and global issues, but he was also heavily involved in national and local problems – sometimes purely technical ones. In addition, Palmstierna moved in several different spheres. Not only was he active as a scientist; he also wrote regularly for *Dagens Nyheter*. Besides, he was an active Social Democrat. This combination of scientific,

---

7 *Miljövårdsforskning. Betänkande del 1. Forskningsområdet* (Stockholm: Ministry of Agriculture, 1967); *Miljövårdsforskning. Betänkande del 2. Organisation och resurser* (Stockholm: Ministry of Agriculture, 1967).
8 Lars J. Lundgren, *Acid Rain on the Agenda: A Picture of a Chain of Events in Sweden, 1966–1968* (Lund: Lund University Press, 1998).

media-based, and political capital gave him a unique platform from which to operate.[9]

The breakthrough of environmental issues in Sweden was intimately intertwined with that of Hans Palmstierna. In the autumn of 1967, he gained recognition as a scientific expert on environmental and future-orientated issues. Paul Warde and Sverker Sörlin have described this knowledge as a special type of scientific meta-expertise. They argue that postwar environmental concepts and scientific expertise about the future were co-produced. In their view, the concept of the environment had a temporal direction right from the start, a direction which pointed ahead to a looming catastrophe. Knowledge about future trends and scientific expertise in the environmental field were developed jointly as two sides of the same coin.[10] But how did this happen in Sweden in the autumn of 1967?

Hans Palmstierna as an alarm clock

On Friday 27 October, Hans Palmstierna was interviewed at considerable length on the TV news. At that time there was only one television channel in Sweden, and the evening news was popular. During the programme, a copy of *Plundring, svält, förgiftning* was displayed. Its cover showed a picture of the tree of knowledge of good and evil. The book was fresh off the press, and Palmstierna was still unknown to most people. He told the reporter about the 'hugely complex poisoning we're being exposed to'. Against a background of pictures of smoking chimneys, sludge pouring from wastewater pipes, and traffic jams, he talked about lead, mercury, and phosphates. He added that famine was a permanent global condition. Within a decade it would hit us. The currently rising meat prices were a harbinger of a world with insufficient food, he said.[11]

That same day, *Plundring, svält, förgiftning* was featured in the country's biggest tabloid, *Expressen* (independent liberal). The paper claimed that the book was one of the most pessimistic to date. The writer asked: 'How long do we really have left on Earth? Ten years? Fifteen?' He emphasized that the most important thing happening

---

9 David Larsson Heidenblad, 'Boken som fick oss att sluta strunta i miljön', *SvD*, 23 October 2017.
10 Warde and Sörlin, 'Expertise for the Future'.
11 'Aktuellt', SVT, 27 October 1967.

right now was that 'we are finally trying to measure the full extent of the catastrophe' and are 'starting to get close to the truth'.[12] A few days later, the Scanian broadsheet *Skånska Dagbladet* described the book as 'the toughest, most concise reckoning imaginable with modern civilization's waste of nature's assets'.[13] The book had an immediate impact. However, it took a couple of weeks for it to move into the main focus of the press.

In an editorial on 11 November, *Dagens Nyheter* discussed both *Plundring, svält, förgiftning* and the report submitted by the government enquiry into natural resources. Hans Palmstierna's book was described as a 'fact-packed and fascinating thriller about the state of the planet and humanity's needs'. Both the book and the report were highly recommended. The two texts could alert people and spur them to act. However, the editorial writer also felt that 'awareness of the environmental problems' was already well under way. Knowledge existed among politicians, industry, and the general public about what was happening and what had to be done. 'Nor is the will to act lacking', the writer stressed. In this way, grave insights into the crisis were combined with a measure of confidence.[14]

On the following day, Palmstierna's book was discussed on the editorial page of the social democratic tabloid *Aftonbladet*. The writer said that the book's author did not hold back in showing 'what an unsustainable development we've ended up in'. In a limited space, Palmstierna had taken a comprehensive approach to 'the gigantic complex of problems on whose solution the future depends'. What made Palmstierna's contribution particularly commendable was that he did not just focus on the problems. According to *Aftonbladet*, he was constantly looking for constructive solutions. His book placed some hope in socialism, scientific enquiry, and international solidarity. 'We have every reason to wish', concluded the editorial writer, 'that his book not only reaches Swedish readers – but also reaches beyond our borders'.[15]

Other voices in the press seized on the apocalyptic elements. In the liberal tabloid *Kvällsposten*, Staffan Ulfstrand wondered if we had come to the beginning of the end. The looming catastrophe

---

12 Ulf Nilsson, 'Hur lång tid har vi kvar på jorden?', *Exp*, 27 November 1967.
13 Ivar Peterson, 'Samhället plundrar våra naturvärden', *Skånska Dagbladet* (*SkD*), 2 November 1967.
14 Anon., 'Sent på jorden', *DN*, 11 November 1967.
15 Bengt Sjögren, 'Internationell planhushållning – ett livsvillkor', *Aftonbladet* (*AB*), 12 November 1967.

would 'hit the whole planet'. There were no new continents to escape to. A few 'lunar and planetary journeys' would not solve the population problems. *Plundring, svält, förgiftning* was presented as insightful, well-documented, and shocking reading. It should be put in everyone's hands.[16] However, Palmstierna's strident tone of voice was a matter of some concern. Many people will have felt that 'Palmstierna is peddling doom completely unnecessarily'.[17] And was it not the case that the 'compelling facts of science' were starting to become 'like ordinary background music?', wondered the cooperative movement's weekly magazine *Vi*.[18]

Most press voices, however, agreed that Hans Palmstierna's book was an important alarm clock. The only critical voice was Nils Landell, writing in the right-wing broadsheet *Svenska Dagbladet*. He argued that the book was permeated by too much pathos and not enough facts. There should have been more examples and less 'irrelevant speculation'. In particular, Landell criticized Palmstierna's political position-taking, expressing his doubts that a socialist government would be best suited to tackle the serious problems. On the contrary, all nations, whatever their social system, should work to solve the global issues.[19] Landell's criticism was of marginal importance, though, and it did not generate any discussion. The dominant opinion in the press was that *Plundring, svält, förgiftning* was interesting, accessible, and scientifically irreproachable. It was an important book which should be read by many.

From knowledge to action

On 21 November, Hans Palmstierna wrote about the report submitted by the government enquiry into natural resources in the article series 'Miljö för framtiden' [Environment for the future]. He began by pointing out that the mercury emissions from the pulp industry had caused mass unemployment in the fishing industry. The biocides threatened higher forms of wildlife. The phosphates in detergents had caused algal bloom in lakes, and the sulphur in the acid rain posed a great danger. All of these problems had come to the public's attention. 'Many surprises are still in store', he wrote.

---

16 Staffan Ulfstrand, 'Början till slutet (?)', *KvP*, 15 November 1967.
17 Bertil Walldén, 'Klockan var mer än vi trodde', *Vestmanlands läns tidning* (*Vlt*), 15 November 1967.
18 Anders Clason, 'Katastrofskval', *Vi*, 18 November 1967.
19 Nils Landell, 'Väckarklocka mot förgiftning', *SvD*, 20 November 1967.

Future government enquiries would probably 'present equally unpleasant revelations'.

How had we put ourselves in this situation? Why had we ruined our environment to the extent that our living conditions were threatened? Palmstierna was clear about the answer: 'We wanted a rapidly rising standard of living.' To make this possible, production had focused on making things as cheaply as possible with no regard to the long-term consequences. Equally catastrophic was the fact that 'we prefer to forget about the goods we have consumed'. By paying the lowest possible price for waste disposal, it had been possible to raise the standard of living very quickly. But nature had not yet presented its bill. 'Will it be so high that we cannot pay it?' he asked.

Palmstierna pointed out that the 1964 government enquiry into natural resources had been commissioned in order to gain an overview of the situation. Its report provided 'an extremely clear and easy-to-read survey' of the nature and seriousness of the problems. The report could be read by anyone who was interested in the issues. Palmstierna stressed that the government enquiry presented new and frightening facts, including Svante Odén's findings about acid rain. It also provided insight into 'the very limited knowledge we so far possess'. To remedy this, target-orientated research and greater research resources were required. State-authority inspections of industries and municipalities should also be intensified.

Palmstierna emphasized that politicians needed comprehensive information from experts in order to make well-considered decisions. The government enquiry into natural resources had now supplied this. What was lacking, though, was a preliminary action programme. Such a programme could complement the information and make it more useful to politicians. The costs should not be a deterrence factor. People's individual standard of living could not be allowed to 'continue to rise at the expense of our future health and our children'.

Palmstierna concluded the article by pointing to an American enquiry which had exposed even more frightening facts than the Swedish one. In that regard, he said, Sweden was fortunate in being ten years behind the United States. However, the American enquiry did make a couple of concrete proposals which he valued. The first was to return consumed material to the production process as much as possible. Developing rational 'circular processes' would make it possible to avoid the worst damage. The second suggestion was to establish an experimental city in which new technology could be tested at the state's expense. Successful innovations and systems could then be spread throughout the country. Such a city could also be used to train environmental conservation experts, a professional group which Palmstierna felt there was a great need to establish.[20]

Palmstierna's article displayed his broad range. There was no doubt about how seriously he regarded the situation, but it was also clear that he perceived considerable scope for action. Palmstierna was concrete and forward-looking. He also showed great faith in politics, technology, and science. This relationship has been highlighted by Jonas Anshelm, who argues that Palmstierna's approach hence did not challenge the Social Democrats' traditional progressivist optimism. This relationship was crucial, Anshelm says, when Palmstierna came to be given a key political role as the Social Democrats' environmental policy was being formulated.[21] In the autumn of 1967, though, Palmstierna did not have such a role; he had not yet been offered any political mandates. His expertise on environmental and future issues was becoming entrenched, however. The decisive factor was that he was perceived as a man of action.

In early December, *Dagens Nyheter* published an extensive and highly appreciative review of *Plundring, svält, förgiftning*. The expressive headline was 'From knowledge to action'. The review was written by the author and engineer Sven Fagerberg. Fagerberg was an influential voice in the Swedish public debate of the 1960s, and he had been discussing crucial global issues for a long time. His review proceeded from the progressivist optimism which he felt had characterized society's leaders during the early postwar period, especially in the technological field. 'The clear advances made in many places seemed to confirm that we were on the right track.' Over time, though, the picture had begun to darken. Question marks were raised about the global use of resources and the direction of development. 'Our prosperity rests on a false foundation', he asserted, 'on a degradation process of a one-way nature.' This meant that we were stealing 'from future generations – our own and, not least, those of developing countries'. This was a 'bitter truth' which political parties and interest groups found difficult to accept because their leaders had been shaped by the early postwar optimism about progress. The new perception of major problems conflicted with

---

20 Hans Palmstierna, 'Vår smutsade värld', *DN*, 21 November 1967.
21 Anshelm, *Socialdemokraterna och miljöfrågan*, p. 18.

The big breakthrough of environmental issues 33

that optimism. The leaders therefore clung to an outdated worldview. Still, Fagerberg said that they did not do this out of ill will, but because of innocence and incompetence.

Nonetheless, the disastrous situation was obvious. 'The problem is', Fagerberg wrote, 'how to make the existing knowledge come alive.' This is what he felt Palmstierna did in such a praiseworthy manner. The book was 'very well supported by facts'; but it was at the same time compelling and action-orientated. Palmstierna's greatest merit was that he belonged 'to the few scientists who feel their responsibility and realize that they must intervene in the practical course of events'. In addition, he 'was constantly indicating ways to take practical action'. This, Fagerberg felt, contained 'a measure of effort, of vitality' which was absolutely necessary. Anyone who wanted to change the world had to take risks. 'If we wait until light has been shone into every dead-end corner of a set of problems, we will be too late.'

Sven Fagerberg's review portrayed Hans Palmstierna as a courageous scientist with a broad orientation and a sense of responsibility. Fagerberg hoped that *Plundring, svält, förgiftning* could make environmental conservation a key political issue. Perhaps Sweden could even become a pioneering nation? In conclusion, he emphasized that the country had a good economy, high technological expertise, and skilled researchers. In addition, no resources were being wasted on military ambitions or space rituals. 'We are practical by nature and think best about material things', he wrote. 'We are not really interested in anything at all and are thus free to become involved.'[22] The time for environmental protection and Hans Palmstierna was now.

Sven Fagerberg's review is one of the clearest examples of how knowledge about a looming environmental crisis and Hans Palmstierna's future-orientated expertise were being co-created in the autumn of 1967. However, Fagerberg was far from being alone in his assessment. On the contrary, at the beginning of December there was great unanimity in the Swedish public arena that Palmstierna was a knowledgeable, pragmatic, and action-focused environmental debater. In the liberal broadsheet *Göteborgs Handels- och Sjöfartstidning*, Bengt Hubendick, one of the most high-profile ecologically orientated voices in Sweden at this time, wrote that *Plundring, svält, förgiftning* should be distributed to 'everyone in a

---

22 Sven Fagerberg, 'Från kunskap till handling', *DN*, 3 December 1967.

position of political and technological responsibility' along with 'demands to read it through and ponder it'.[23] In the likewise liberal broadsheet *Göteborgsposten*, Göran Michanek stated that Palmstierna had achieved something new. He had shown that environmental destruction was not a far-off threat of disaster: it concerned us and our children.[24] It was high time to move from knowledge to action.

Science, politics, and the limits of expertise

In December, the widespread support for Hans Palmstierna and his book *Plundring, svält, förgiftning* contrasted with the intensive debate that arose over *Människans villkor*. The two debate books were outwardly very similar, but they came to be perceived in very different ways. The ensuing pages review the reasons why this happened and what the consequences were.

On Thursday 7 December, the day before *Människans villkor* reached the bookshops, it was featured in the televised weekly magazine *Monitor*. The broadcast began with three terms rolling past on the screen: global fire, global famine, global poisoning. Then pictures were shown of starving, emaciated children from the developing world. The powerful images were ironically accompanied by a sung version of 'God who holds the children dear', the most widespread evening prayer for children in 1960s Sweden.

The unsettling opening scene of the programme was followed by Georg Borgström, filmed in his office and surrounded by books, talking about global injustices, malnutrition, and overpopulation. He emphasized that we were in the 'initial stages of a monumental crisis' and that we must all take off our blinkers. 'An unpleasant reminder in the midst of the early Christmas rush, isn't it?' said the narrator, whereupon photos of chimneys, car-exhaust emissions, and polluted watercourses were shown. 'Pictures like these also arouse discomfort', continued the voice; 'we are pouring toxins and gases and dangerous substances into nature and over ourselves, with consequences that we know far too little about. Only that they may be devastating.'

Monitor continued by interviewing several of the researchers behind the book. Ecologist Bengt Lundholm, who had been the secretary of the 1964 government enquiry into natural resources,

---

23 Bengt Hubendick, 'Nu är det allvar', *Göteborgs Handels- och Sjöfartstidning* (*GHT*), 4 December 1967.
24 Göran Michanek, 'Väckarklocka med skräll', *GP*, 29 November 1967.

The big breakthrough of environmental issues

talked about DDT and mercury. Physicist Tor Ragnar Gerholm focused attention on the world's nuclear weapons. Carl-Göran Hedén, professor of bacteriology at the Karolinska Institute, criticized the prevailing political system. He felt that the national organization and the short legislative terms were a fragile foundation on which to stand in a situation where the survival of the human species was at stake.

The programme concluded with a studio debate between Lars Gyllensten and the Social Democratic government minister Krister Wickman. In this section, the order was reversed. It was not the scientist Gyllensten who held the politician Wickman to account but vice versa. The government minister spoke first. He rejected the image of politicians painted in the book. The researchers seemed to believe that politicians 'are cynically exploiting an easily led, ignorant mass of voters'. This view revealed a deep contempt for politicians but, even more seriously, a contempt for voters. Did Gyllensten want to replace political democracy with rule by technocratic experts?

The debaters' body language was significant. Wickman sat leaning forward in an assertive position, whereas Gyllensten looked down at the floor. He averred that he was not attacking democracy, nor did he despise politicians and voters. All he wanted was greater scientific influence. Wickman then steered the conversation on to the topic of long- and short-term goals. The two men discussed the role played by politicians in creating public opinion. On this point there was considerable agreement. Both felt it was the politician's task to lead and shape opinions, not merely to implement what was possible at any particular time. Wickman ended by underlining that the whole problem was to a large extent a 'matter of knowledge and awareness'. He conceded that for a long time society had underestimated the risks in the environmental field. But right now 'we are experiencing a really noticeable change in attitude about these issues'. The conditions for reaching a solution were good. In this way, *Monitor* came to a mildly optimistic and reassuring conclusion.[25]

The next day, the press reported on both the television programme and the book release.[26] *Svenska Dagbladet* also published an initial review. It characterized the scientists' initiative as commendable in

---

25 'Monitor', SVT, 7 December 1967.
26 Rune Johansson, 'Hotet mot mänskligheten', *DN*, 8 December 1967; Gall, 'Syndafloden som stiger', *SvD*, 8 December 1967.

principle but not very constructive in practice. Their book contained no suggestions for 'concrete political measures' but only 'well-meaning and vague prescriptions'. The reviewer wondered whether 'a few concrete instructions' might have been given to politicians instead of 'just generally scolding them and declaring them out of date?' Such an approach might have assisted in the building of a willingness to cooperate.[27]

A few days later, *Aftonbladet*'s editorial page continued along the same lines. It welcomed scientists intervening in the public debate. But it strongly condemned the contempt for politicians – and ultimately for voters – expressed in the book. Carl-Göran Hedén and Lars Gyllensten were particularly singled out for criticism. They seemed to regard scientists as enlightened truth-seekers and politicians as power-hungry deceivers. The editorial writer stressed that this type of contempt was a 'fruitless starting point' for establishing greater cooperation. In addition, the researchers 'had cheerfully helped to create the technological advances which now constitute deadly threats to humanity'. It was therefore an 'unusually unjustifiable arrogance' to portray scientists, as opposed to politicians, as 'moral clean-living types'.[28]

Criticism was harsh on *Dagens Nyheter*'s editorial page as well. The book's subtitle in itself prompted questions. Why were scientists only addressing politicians? Did the big questions of the future not concern everyone? The editorial writer stressed that societies neither could nor should be led by 'hierarchical elite networks of researchers-politicians-engineers'. In addition, a lot was beginning to happen in the world. Both in Sweden and abroad, people were starting to take the environmental dangers more and more seriously. While important to this development, politicians and researchers 'were not more central than other influencers and power factors in society'.[29]

Of the twelve scientists behind *Människans villkor*, only Carl-Göran Hedén tried to respond to the criticism. He said that he considered politicians to be a great resource but that party politics posed a serious danger. As far as possible, it should be replaced by 'the scientific method'. For Hedén, it was not a question of whether scientists or politicians were the morally superior category. It was about different ways of working. He preferred the scientific method and argued that it should have more influence on how society was

---

27 Thure Stenström, 'Naturvetarna och världens nöd', *SvD*, 8 December 1967.
28 Anon., 'Forskare diskuterar politik', *AB*, 12 December 1967.
29 Anon., 'Vetenskap och politik', *DN*, 11 December 1967.

governed. Hedén praised 'dynamic real democracy', by which he meant frequent referendums on specific issues.[30]

Hedén's contribution, however, did not pour any oil on the troubled waters. Throughout the month of December, *Människans villkor* continued to be criticized in the press. There was widespread insistence that the scientists raised important issues, but also that they evinced elitist and anti-democratic tendencies.[31] The Liberal Party politician Carl Tham (later a Social Democrat) was especially censorious. He argued that the researchers' attack on the politicians had a 'disquieting kinship' with the criticism of 'the principles of democracy previously asserted by the far right'.[32] In this context, it is noteworthy that the book's editor, Karl-Erik Fichtelius, did not participate in the debate. This may seem surprising, as Fichtelius was an experienced debater; but the reason is that he was in the US on a lengthy stay as a visiting researcher.[33]

The extensive and unanimous criticism of *Människans villkor* shows that there were sharp limits on scientific expertise in 1960s Sweden. Researchers were welcome to define problems, spread knowledge, and shape opinions. But when they moved into the field of political decision-making in a confrontational manner, they encountered strong opposition. At the same time, the attention paid to *Människans villkor* undoubtedly helped to circulate knowledge about a global environmental crisis among the general public. Both in the book and in the press debate, it became clear that many people had begun to regard environmental destruction as a connected set of problems intimately linked with other issues of survival. However, scientific expertise about the future circulated in an ambivalent manner. Some researchers, such as Lars Gyllensten and Carl-Göran Hedén, were viewed with scepticism and suspicion. By contrast, Hans Palmstierna would further strengthen his own position.

---

30 Carl-Göran Hedén, 'Ett genmäle om vetenskap och politik', *AB*, 19 December 1967.
31 Jean Braconier, 'Utmaning till politikerna', *Sydsvenska Dagbladet* (*SDS*), 15 December 1967; Folke Johansson, 'Vetenskapsmän och politik', *Upsala Nya Tidning* (*UNT*), 21 December 1967; Anon., 'Politik och vetenskap', *KvP*, 30 December 1967; Erik Hjalmar Linder, 'Debatternas år', *GP*, 31 December 1967.
32 Carl Tham, 'Forskare och politiker', *DN*, 20 December 1967.
33 Letter from Karl-Erik Fichtelius to Per Gedin, 5 January 1968, vol. 106, Albert Bonniers förlag II (publisher's archive), Centre for Business History.

The consolidation of Hans Palmstierna's expertise

In the lively debate about *Människans villkor*, reference was often made to *Plundring, svält, förgiftning*. However, the growing criticism against scientists was never levelled at Palmstierna. On the contrary, his book continued to be praised. On 13 December, the liberal broadsheet *Sydsvenska Dagbladet* singled it out as one of the 'most acerbic, most ingenious, best informed, and best presented' debate contributions made in Sweden in a very long time. It was underlined that Palmstierna had a broad education, not only in the natural sciences but also in the humanities. He was able to go beyond his own narrow area of expertise and dared to comment on the really big issues. 'And this is surely what is necessary', wrote the reviewer, 'that there is an elite of fearless debaters with a broad enough frame of reference to be able to think in an interdisciplinary way'.[34]

The emphasis on Palmstierna's great breadth and wide-ranging knowledge was typical of how his expertise circulated. Another key aspect was that he was characterized as an optimist. In the words of one writer, people might have presumed that 'Palmstierna with all his knowledge' would long ago have stopped believing in the value of appealing to 'individual or collective reason'. But this was not the case. Although he saw the seriousness of the situation, Palmstierna was an optimist who believed in people. Joint efforts could 'avert the impending misfortunes'.[35]

On Christmas Eve 1967, Gösta Bringmark wrote a column in the largest social democratic broadsheet, *Arbetet*, based on the biblical story of the expulsion from the Garden of Eden. Bringmark said that modern humans had now learned that 'knowledge really is both good and bad'. He underlined that 'our technology is killing our own existence' and that 'humanity is a diseased organism in nature or a parasite on the Earth'. The speed at which these views had become established was astounding, he said. Environmental issues had 'finally begun to break through on a broad front'. Even so, Bringmark was concerned that a rift between scientists and politicians was being created. 'It is of the utmost importance that this chasm does not deepen', he urged. The person who guaranteed that it would not was Hans Palmstierna. Better than anyone else,

---

34 Lars Holmberg, 'Giftvatten, snuskland', *SDS*, 13 December 1967.
35 Erik Nyhlén, 'Vår nedsmutsade värld', *Borlänge Tidning* (*BoT*), 18 December 1967.

he had been able to take a 'concise and motivating approach' to the momentous matters of destiny.[36]

The contrasts in how the Swedish press handled the contributions of the various scientists were striking. The knowledge they cited was largely the same; but the way in which the researchers' expertise circulated differed. Palmstierna was presented as reasonable and politically concrete, whereas the researchers behind *Människans villkor* were regarded as arrogant and vague. This contributed to a further strengthening of Palmstierna's position. Among a growing chorus of scientific warning voices in the late autumn of 1967, he stood out as sensible and pragmatic.

In addition, Palmstierna had access to a key media platform: *Dagens Nyheter*'s culture page. In the 1960s, the newspaper held a leading position in the Swedish public debate and was a driving force behind getting environmental issues onto the agenda, not least through Barbro Soller's journalism. Hired as a general reporter for the newspaper in 1964, she gradually developed into an environmental reporter. Soller became Sweden's, and one of the world's, first full-time and on-staff environmental journalists.[37] Besides, *Dagens Nyheter* made several efforts to bring in scientists as writers in the newspaper. The article series 'Miljö för framtiden' was crucial in this respect. Only one scientist was given the opportunity to write two articles for it: Hans Palmstierna.

On 29 December, 'Insikt, kunskap, handling' [Insight, knowledge, action] was published. It was the ninth and final part of the article series. Palmstierna began by stating that 'the realization that the Earth is small, and that humanity has the power to destroy its own possibilities of continuing to live, dawned late'. This realization, he argued, had begun to take shape when the atomic bombs fell on Hiroshima and Nagasaki, but no 'mass movement' had ever arisen. Instead, people became used to 'living under the threat of annihilation'. In the shadow of the bomb, however, new insights had emerged. Researchers had gradually discovered that many industrial processes were highly risky for 'the nature we live in and live off'. The environment was more sensitive than we had thought.

Palmstierna stressed that we have now 'brutally experienced that the Earth *is* small and life fragile'. These insights were no longer

---

36 Gösta Bringmark, 'Människan som parasit eller Kunskapens träd på gott och ont', *Arbetet (Arbt)*, 24 December 1967.
37 Djerf Pierre, *Gröna nyheter*; Larsson Heidenblad, 'The Emergence of Environmental Journalism'.

'the property of a small minority', he wrote, 'but now belong to the general public'. However, it had taken a long time to reach this crucial point. The scientists, who had sensed the risks intuitively for a long time, had not had 'enough evidence to be able to convince'. Palmstierna claimed that society had 'demanded too much detailed knowledge before it wanted to believe the warnings'. Many people had reacted unfavourably to the scientists' insights and rejected them.

The reason, Palmstierna said, was that the lines of communication between scientists and politicians had not worked. 'There is still no calm and trust-based dialogue', he wrote. For that reason, 'a new group of interpreters must be singled out from among the ranks of the scientists'. These people would be able to 'translate the findings and warnings of science into clear and distinct normal prose, so that [...] the authorities and the general public can acquire a true picture of what is happening'. This was the role he took upon himself. It was the first necessary step towards having a viable environment. The aim was to create readiness among politicians and the general public to accept the intrusive and costly measures that the situation required. Nor could these information efforts stop at national borders. The problems were transnational, and so the scientists' opinion-building work had to be transnational too.

In conclusion, Palmstierna again stressed that the individual standard of living could not be allowed to rise further at the expense of the shared environment. If this continued, we would soon no longer have 'any viable environment to live in – and live off'. He underlined that in this serious situation, optimism about progress was not warranted; but nor was pessimism. 'We must [...] face the facts', he wrote, 'and act rationally on the basis of the knowledge we possess.' Only in this way could we guarantee that our generation, as well as future ones, 'will survive in a manner compatible with human dignity'.[38]

Palmstierna's article concluded the Swedish environmental debate of 1967. As a result of his and other scientists' actions, that debate had changed fundamentally in a brief period of time. Knowledge of a global environmental crisis and scientific expertise regarding the future were now circulating intensively in the Swedish public sphere. However, a social knowledge breakthrough cannot be fully studied by examining the public sphere alone. What happens

---

38 Hans Palmstierna, 'Insikt, kunskap, handling', *DN*, 29 December 1967.

The big breakthrough of environmental issues

there is important; but there are other arenas which are also decisive for how knowledge moves and operates within a society. In the following section, I will therefore shed light on the breakthrough of environmental issues on the basis of two meetings held in the corridors of power.

Forskningsberedningen's meeting on environmental conservation issues

On Monday 4 December 1967, *Forskningsberedningen* (the advisory council on research), chaired by Sweden's prime minister Tage Erlander, met to discuss research on environmental conservation. Despite its name – *beredning* in Swedish has connotations of preparation and drafting, especially in a legislative context – *Forskningsberedningen* was not a preparatory body. It functioned as a meeting place for politicians, scientists, and other key figures in government and industry, having been set up in 1962 in order to deepen cooperation between them. The starting point was that research was thought to play a key role in the development of society. *Forskningsberedningen* was a step towards a more active government research policy. Its significance was highlighted by the fact that the Swedish prime minister himself chaired its work.[39]

The meeting on 4 December brought together forty-seven people, including a large number of government ministers, professors, and directors of various authorities. In addition to the regular participants in *Forskningsberedningen*, fourteen individuals had been summoned especially to attend on this particular occasion. They included Hans Palmstierna and Svante Odén. *Forskningsberedningen* undoubtedly assembled a social elite, and the question is how knowledge and expertise circulated in such a context.

The meeting began with a long speech by Prime Minister Erlander. He said it was now obvious that 'our environment is seriously threatened'. The warning signals were coming more and more often, both in Sweden and abroad. Lakes were eutrophying, acid rain was

---

[39] Peter Stevrin, *Den samhällsstyrda forskningen: En samhällsorganisatorisk studie av den sektoriella forskningspolitikens framväxt och tillämpning i Sverige* (Stockholm: Liber, 1978); Rune Premfors, *Svensk forskningspolitik* (Lund: Studentlitteratur, 1986); Tunlid and Widmalm (eds), *Det forskningspolitiska laboratoriet*; Per Lundin, *Lantbrukshögskolan och reformerna: Från utbildningsinstitut till modernt forskningsuniversitet* (Uppsala: Swedish University of Agricultural Sciences, 2017), pp. 109–110.

falling from the sky, and fish were becoming contaminated. The ongoing poisoning was worsening the conditions for agriculture and forestry. Plant and animal life was being threatened. 'Are we on the way', the prime minister wondered, 'towards gradually, and partly imperceptibly, making our existence impossible through environmental destruction?'

Erlander continued his speech with a historical review. He emphasized that 'in a situation of highly obvious material shortages, the environment comes in second place'. Because this had previously been the case, the overarching political goal had been said to consist in 'creating increased production, creating jobs, rapidly raising the standard of living'. As a result, the environment in Sweden, as in other industrialized countries, had been exploited to secure 'basic material needs'. Erlander emphasized that 'not for one moment' should anyone underestimate what this technological and economic development had meant to people. It had enabled 'increased consumption, greater social security, more leisure time, and a number of other [good] things'. This development would continue to be safeguarded, but the environment must be 'taken into account in a completely different way than it used to be in the past'.

Erlander proposed that environmental issues should be perceived as 'part of the social reality experienced by the individual' to a greater degree than before. As the standard of living increased, so did the demands on the environment. 'Stemming the destruction of nature, pollution, and poisoning' would ensure security. Achieving this aim called for cooperation between the central government, municipalities, and the world of business. It was unreasonable, Erlander felt, to demand that individual companies be able to perceive the long-term environmental consequences of their own operations. Environmental problems were an issue for society as a whole. It was at the political level that rules and boundaries had to be set. This required 'a shared sense of values' and acting in solidarity.

Towards the end of his speech, the prime minister adopted an existential tone and looked ahead. He acknowledged that it was 'easy to feel powerless when faced with the dimensions of the environmental problems', but said that people did not have to feel that way. 'We have greater economic resources than ever', he pointed out, adding that 'researchers have made pioneering efforts to make us aware of the urgency of the environmental problems.' Of course, greater knowledge and a balanced overview were required; but 'ultimately everything depends on what we are willing to do'. Erlander stated that the government was prepared to go further, adding that

The big breakthrough of environmental issues 43

'the distance between knowledge and action must be as short as possible'. New research and new technology would be developed. 'What today's discussion is about', he concluded, 'is how, with the help of research, we can secure a viable environment for humanity, an environment that we can pass on to future generations.'[40]

Erlander's speech shows that, at that time, environmental issues had taken root at Sweden's highest political level. The prime minister was fully aware of their seriousness. His insistence that a rapidly rising standard of living had had unforeseen consequences echoed the description of events by Palmstierna and other debaters. One important difference, though, was that Erlander was very careful to emphasize the favourable aspects of development as well. People had become better off. The material advances were real and desirable. The clock would not be turned back. It was his hope that research and politics could deal with the unwanted side effects and ensure continued positive social development.

The subsequent discussion was based on the report of the government enquiry into natural resources. In particular, organizational issues were debated. The enquiry had proposed that an advisory council on environmental issues should be established under the direct leadership of the minister for agriculture. This council would be mandated to fund both basic research and goal-orientated research. Many of the professors who were present welcomed this proposal. They argued that research needed more resources if it was to be able to change society. At present, a shortage of researchers and insufficiently attractive project positions posed obstacles to success. If the government was serious about its new focus, the basic grants had to be increased. It was not possible to either attract or retain the best researchers with one- and two-year contracts.[41]

The organizational relationship between the new National Environment Protection Board and the proposed Environmental Advisory Council [miljövårdsberedningen] was the subject of intense debate. Those participants in the meeting who were doing active

---

40 The minutes of the meeting are among Birgitta Odén's bequeathed documents, which are being stored at the Department of History in Lund while awaiting formal archiving. They consist of two binders, referred to in this book as BO 1 and BO 2. BO 1, 'Statsministerns anförande vid Forskningsberedningens sammanträde den 4 december 1967' [The prime minister's speech at the meeting of *Forskningsberedningen* on 4 December 1967], pp. 1–5.
41 BO 1, 'Mötesprotokoll från forskningsberedningen 4 december 1967', pp. 4–5, 7–9, 12, 14–15.

research stressed the value of independence while other voices argued that a coordinated approach could shorten the step from research to action. Lars Brising, who was director-general of engineering in the Swedish Air Force, said that 'there was everything to gain' from making the National Environment Protection Board an 'expert and thereby forceful executive institution'.[42] The value of coordination was also emphasized by Martin Fehrm, director-general of the Swedish National Defence Research Institute (FOA). He stressed the importance of 'utilizing the results of research when undertaking social planning', adding that it was important for existing knowledge to be fitted into 'a model plan or overall picture' as soon as possible. According to Fehrm, this type of systems analysis was 'one of the most important elements of goal-orientated environmental conservation research'. He also underlined that the cost estimates supplied by the government enquiry into natural resources were at the lower end. If the government was serious about 'attacking the natural-resource problem', it should be prepared for 'significantly higher costs'.[43]

The economic aspects were also highlighted by Erik Dahmén, professor of economics at the Stockholm School of Economics. He said that there were 'very strong socioeconomic reasons' for taking measures against environmental degradation and lending greater depth to environmental research. Dahmén felt that any 'sacrifice of environmental values' needed to be perceived as costs – just like 'raw materials, capital, and labour'. Currently there was no price mechanism in operation, which meant that 'short-term consumption preferences' were favoured at the expense of environmental values.[44] Dahmén would further develop these ideas the following year in his high-profile debate book *Sätt pris på miljön: Samhällsekonomiska argument i miljöpolitiken* [Put a price on the environment: socio-economic arguments in environmental policy] (1968).[45]

Bank director Tore Browaldh continued along the same lines as Dahmén. He drew attention to the possibility of 'solving the environmental problems in the long run by utilizing market-price formation'. This could be done, for example, by putting a special tax on environmentally unfriendly products. However, Browaldh stressed

---

42 *Ibid.*, p. 8.
43 *Ibid.*, p. 6.
44 *Ibid.*, p. 9.
45 Erik Dahmén, *Sätt pris på miljön: Samhällsekonomiska argument i miljöpolitiken* (Stockholm: SNS, 1968).

that a tax policy specific to Sweden risked doing hefty damage to Swedish companies which operated in a highly competitive international market. Even so, he suggested that Dahmén should be mandated to lead a working group that could develop a proposal as to how market-price formation might be designed and introduced in the environmental field. Browaldh further envisioned that a powerful research effort in the environmental field could in due course lead to the creation of a new Swedish industrial sector in environmental technology.

Another aspect that was repeatedly mentioned was the need for international cooperation. On this subject, there was widespread agreement but also some concern. Sune Bergström, professor of medical and physiological chemistry at Lund University, said that there was a great lack of knowledge and interest within the Organisation for Economic Co-operation and Development (OECD). Only 'a few member states' were 'aware of the seriousness of the problem'. For this reason, Sweden could not passively await future international agreements. Instead, we should strive to become 'a pioneering nation and, through this primary effort, promote interest in the issue in other countries'.[46] Nobel Laureate Arne Tiselius, professor of biochemistry at Uppsala University, also felt that the government should be proactive. He suggested that Sweden should encourage the establishment of an international research council which might assume strategic responsibility for initiating 'urgent targeted research'.[47]

Towards the end of the meeting, Hans Palmstierna spoke. Stressing 'the need to inform the public', he also argued that biologists should be employed 'to a significantly greater extent' in order to inform engineers. Palmstierna agreed with Dahmén's comment that 'excessive consumption benefits' had been extracted, with 'destroyed nature as the result'. However, he emphasized that the distribution was uneven. The one who destroyed nature and profited from it was seldom the one who was adversely affected as a result. There was an injustice here that must not be forgotten.[48]

Clearly, then, the *Forskningsberedningen* meeting had significant points in common with the public environmental debate. Knowledge and expertise circulated in similar, though not identical, ways. The main difference was that this meeting – with the exception of

---

46 BO 1, 'Mötesprotokoll från forskningsberedningen 4 december 1967', p. 11.
47 Ibid., p. 14.
48 Ibid., p. 15.

Tage Erlander's introductory speech – did not talk about environmental issues in terms of humanity's survival. It is also striking that overpopulation and global injustices were not discussed. At this meeting, the environmental problem was more narrowly defined. There was no talk of a crisis or a future global catastrophe. The discussion was primarily coloured by the report of the government enquiry into natural resources and the more low-key manner in which that report discussed environmental issues. The apocalyptic framework, much to the fore in the public debate, was not important here.

A meeting at the Swedish National Defence Research Institute (FOA)

On 27 November, a week before the meeting of *Forskningsberedningen*, another meeting was held at FOA. That meeting also addressed the question of how research might contribute to the development of society. The focus, however, was not on scientific and technical expertise but on the social sciences and humanities. This was because director-general Martin Fehrm felt that the environmental problems were ultimately bound up with human actions and political decisions. Consequently, scientific and technical knowledge was not enough. In order to give politicians a sufficient basis for decision-making, other skills were required as well.[49]

In May 1967, Martin Fehrm had convened an initial meeting at FOA. Among those invited were three professors: economist Assar Lindbeck, political scientist Pär-Erik Back, and historian Birgitta Odén. At this first meeting, they made it clear to Fehrm that the knowledge he requested did not exist. Researchers had not previously been interested in the historical, political, and economic dimensions of environmental issues. New research efforts were required to produce this knowledge. As a result, the concept of a joint research programme was born.

Another issue that was raised was how to make politicians and the general public realize the seriousness of the situation. Fehrm himself regarded the destruction of the environment as a security threat, comparable to other external threats to society's continued existence. The other participants supported this view. The discussions resulted in the idea of writing a joint debate book. Birgitta Odén

---

49 What follows is based on Larsson Heidenblad, 'Miljöhumaniora på 1960-talet?', pp. 44–45.

was mandated to draw up the guidelines for such a book. On 27 November she presented a discussion paper.

The form she envisioned was 'a modest publication, as accessibly written as possible. In other words: a serious pamphlet, intended to provoke discussion.' In order to have an impact it needed to be published with all due despatch, preferably as early as the spring of 1968. The idea was that experts would speak up and awaken politicians and the general public to an understanding of the *seriousness* of the issue and the necessity of rapid *targeted research*'. Her memorandum, totalling five pages, contained a detailed synopsis which included a list of suggested writers. She placed special emphasis on the preface, which should be written by 'a person who was heeded by public opinion'. The theme should be 'the unintended consequences of the development of prosperity and technology and their disastrous consequences for the future of our children'. The person to whom she wished to entrust this task was the Social Democratic politician and diplomat Alva Myrdal, who worked with disarmament issues at an international level.[50]

The preface would be followed by an introduction containing the group's joint programme statement. This would assert that environmental destruction was 'such a serious threat to our future prosperity that it can be equated with a military security risk'. It was this circumstance that justified FOA's assuming the leadership of the operation. The aim of the research would be 'to acquire better information in order to guide the people who are making the crucial decisions'. The proposed author was Martin Fehrm, with the participation of the whole group.[51]

The ensuing three chapters would examine historical examples of disastrous environmental destruction, political decision-making processes in the environmental field, and the issue of how scientific information was disseminated within political bodies. Birgitta Odén and Pär-Erik Back could assume special responsibility for these parts of the publication. Six chapters of a scientific, medical, and security nature would follow. These would include one chapter by Svante Odén on the acidification caused by precipitation, one by Hans Palmstierna on public-health issues, and one by FOA chief engineer Erik Moberg on the Baltic Sea and Sweden's security policy. The purpose of the last-mentioned chapter was to show that 'the pollution of the Baltic Sea may lead to the Soviet Union making demands on

---

50 Birgitta Odén, 'PM 1' ['Memorandum 1'], November 1967, BO 1, p. 1.
51 *Ibid.*, pp. 1–2.

us, demands which we may find difficult to meet'. The question, however, was whether this could be said openly or whether it should merely be implied.[52] The Cold War context – in which Sweden sought to maintain a neutral position – was much in evidence here.[53]

These chapters, with their concrete problem descriptions, would be followed by a chapter which addressed environmental destruction from a national-economic perspective. This chapter could show that the cost calculations for basically all industrial production were too low 'if the cost of restoring nature is not also included in the calculation'. The chapter would culminate in a plea for 'realistic cost calculation' and a discussion of '*where* the costs of environmental restoration should be taken from'. The thirteenth and final chapter would outline a comprehensive approach to the environmental issues. In this chapter, Martin Fehrm's task would be to emphasize how the ongoing environmental destruction was having adverse effects on human beings in virtually all areas. He would especially highlight the economic, health, and security-policy dimensions of the threats. The key word was 'coordination' of both the research efforts and the political decisions.[54]

Nothing, however, came of these plans for a joint publication. At the top of one of the copies of her memorandum notes, Birgitta Odén has briefly written 'rejected'.[55] Possible explanations for this can be found in the document entitled 'Min föredragning' [My presentation]. In this document, Odén herself raised the question of whether a joint publication was in fact necessary. 'Or has the situation changed after the *DN* debate, Palmstierna's book, the report from the government enquiry into natural resources, and the actions of the National Environment Protection Board?'[56] Her questions

---

52 *Ibid.*, pp. 2–3.
53 In recent years, environmental history research has increasingly drawn attention to the significance of the Cold War context. See Ronald Doel, 'Constituting the Postwar Earth Sciences: The Military's Influence on the Environmental Sciences in the USA after 1945', *Social Studies of Science* 33.5 (2003); John R. McNeill and Corinna R. Unger (eds), *Environmental Histories of the Cold War* (Washington, DC: German Historical Institute, 2010); Jacob Darwin Hamblin, *Arming Mother Nature: The Birth of Catastrophic Environmentalism* (Oxford: Oxford University Press, 2013); Joshua P. Howe, *Behind the Curve: Science and the Politics of Global Warming* (Seattle: University of Washington Press, 2014).
54 Birgitta Odén, 'PM 1', November 1967, BO 1, pp. 1, 4–5.
55 Birgitta Odén, 'PM 1' (copy with margin notes), November 1967, BO 1, p. 1.
56 Birgitta Odén, 'Min föredragning', November 1967, BO 1.

testify to the relevance of the matters raised in this chapter. Between May and November 1967, the Swedish environmental debate had undergone a fundamental change. Knowledge and crisis insights no longer circulated only in specific circles, such as those at FOA, but were moving with great intensity within the public sphere. Politicians and the general public had woken up.

But how much can we really know about the latter category? Is it possible to study the breakthrough of environmental issues in the autumn of 1967 from the perspective of the Swedish general public? What traces remain of those people who did not belong to a social elite which expressed itself in print? It is not easy to answer these questions satisfactorily. However, Hans Palmstierna's rich personal archive does contain a few examples. The earliest I have found comes from an unknown Gothenburg resident named Sören Gunnarsson. In October 1967 he contacted Palmstierna, initiating a correspondence. The following section takes a closer look at this correspondence with a view to shedding light on ways in which the environmental debate could intervene in a layperson's life.[57]

The layperson's voice

Sören Gunnarsson began his first letter by saying that Palmstierna's articles in *Dagens Nyheter* had 'meant a lot to me and stimulated my thoughts to focus on the serious problems you are writing about'. Gunnarsson added that the debate on 'contamination and exploitation of the Earth' had intensified in the past year. He expressed growing unease and concern at 'the ruthlessness with which the big industries are ruining future life opportunities'. This short-term thinking was almost as 'disastrous and challenging' as the exploitation of the Third World. Concerned, he turned to Palmstierna for information and guidance. What was being done by those in charge? What were the people who realized that humanity was threatened actually doing? What could a layperson do other than just read the publications of scientists? Were there any pressure groups or petitions? Could not an office be set up to supply the media with information and debate articles?[58]

---

57 What follows is based on Larsson Heidenblad, 'Överlevnadsdebattörerna'.
58 Letter from Sören Gunnarsson to Hans Palmstierna, undated October 1967, 452/3/2, Hans Palmstierna's personal archive (HP), Swedish Labour Movement's Archives and Library (ARBARK).

Palmstierna replied to Gunnarsson that the extent of the problems and humanity's short-term covetousness worried him deeply. At the same time, though, he was not disheartened. 'Had I not seen a glimmer of hope, I would not have written the book that was published last week: Plundring-Svält-Förgiftning. It attempts to attack the problems, and, where I see solutions, to suggest some possibilities.' Palmstierna urged Gunnarsson to stand up and work for change. 'The fastest way is probably via the political parties. I am trying to do what I can within the Social Democrat movement, since that party is the most conscientious one where these issues are concerned.' Palmstierna also pointed out that, within the Environmental Advisory Council, an ecology committee had been formed which was intended to function as a lobby group. He concluded by agreeing with Gunnarsson's criticism of the 'selfish desire for profit'. This was what was forcing the destruction of the shared environment. 'There should be no owners', wrote Palmstierna, 'in the sense that a person is allowed to destroy their own property so that third parties or future generations will suffer. There should only be stewards.'[59]

Gunnarsson replied on a postcard. He rejoiced that there was 'a book on the way' and hoped it would be a success. 'And that it [will] function better than other "alarm clocks" during this unique autumn.'[60] Shortly afterwards, he sent Palmstierna a newspaper article in which some engineers were interviewed. He said that in itself the article was nothing remarkable, but that it demonstrated the frightening 'cluelessness' displayed by the engineers in the field of environment and energy.[61]

Hans Palmstierna thanked Gunnarsson for 'the naive article'. As he understood it, though, its content was merely a call for 'a more skilfully managed planning of society and more stringent control'. He pointed out that what the engineers had said had been 'filtered through the journalist's feeble intellect' and underlined that bad journalists 'love to end an article with a stupid, preferably derogatory, closing remark'. This is especially true when they have not 'been able to follow along with the conversation'. Possibly 'the discussion by these engineers had made a lot of sense'.

---

59 Letter from Hans Palmstierna to Sören Gunnarsson, undated October 1967, 452/3/2 (HP ARBARK).
60 Postcard from Sören Gunnarsson to Hans Palmstierna, undated October 1967, 452/3/2 (HP ARBARK).
61 Letter from Sören Gunnarsson to Hans Palmstierna, undated October 1967, 452/3/2 (HP ARBARK).

The article prompted Palmstierna to develop his thoughts about the conditions of communication. He felt that a key issue was how to spread these crisis insights to a larger number of people. He distanced himself from those who believed that information did not penetrate people's minds unless it was 'as brainless as a comic strip in the Sunday supplement'. This view was an expression of arrogance, he said, and also totally wrong. 'I have observed the opposite. Good information hits home', on condition that it was not presented in an 'obtuse and offensive manner' but in a 'considerate and wise way'.[62]

Palmstierna and Gunnarsson continued to exchange letters during the autumn. Unfortunately, not all of Gunnarsson's letters are preserved. On the basis of Palmstierna's replies, however, it appears that a couple of letters written in November discussed two environmental issues which were current at that time on Sweden's west coast. The first one concerned the city of Gothenburg's wastewater, which threatened the Göta River and the northern parts of the archipelago. Palmstierna argued that a new treatment and sewage system should be built, which would make 'the sewers' materials useful again for forestry and agriculture'.

The second issue concerned the plans to build a sulphate factory in Väröbacka, just north of Varberg. At that time, there were also plans to build a nuclear power plant in that location. On this issue, however, Palmstierna was not so critical of central planning. He explained to Gunnarsson that there were great advantages to industries being densely located: if they were, the surroundings could not be ruined 'without making it impossible for people to work'. This put pressure on decision-makers to develop cleaner processes. Besides, such a concentration made it easier for society to oversee and control the industries.[63]

As well as discussing the political issues of the day, the correspondence mainly focused on how the environmental struggle should be organized and strengthened. In his postcard Gunnarsson said that pressure groups should be set up, a view in which Palmstierna concurred. The latter noted that scientific groupings were being formed, but he felt that a 'lay committee' would be needed too. As a successful example of this, he pointed to the Scientist and Citizen

---

62 Letter from Hans Palmstierna to Sören Gunnarsson, 6 November 1967, 452/3/2 (HP ARBARK).
63 Letter from Hans Palmstierna to Sören Gunnarsson, 28 November 1967, 452/3/2 (HP ARBARK).

group in St Louis, which 'from a modest beginning has become very influential'.[64]

Palmstierna admitted that he sometimes 'felt very much alone' and that he 'greatly longed for a fighting group'. However, he said that since the publication of *Plundring, svält, förgiftning*, it had become apparent that 'there are many friends'. On '[his] list' were both scientists and administrators. They were all socially aware and consciously or unconsciously left wing. Palmstierna singled out the poet and author Svante Foerster as 'a good and combative person'. He added that he had good contacts with the Stockholm-based group called the Young Philosophers. 'We can form a fighting group like that one and also like the American St Louis group.'[65]

At the end of the letter, in response to a complaint by Gunnarsson, Palmstierna commented on the publisher's pricing of *Plundring, svält, förgiftning*. He stressed that he had not written the book in order to make money, and that the high price (approximately 20 euros in today's value) was only due to the publisher's not believing that it would sell particularly well. However, the first edition of the book had already sold out. 'It can only mean that there are many people who think like us – for the most part.' Palmstierna looked ahead with optimism. 'Let us create a popular movement. The time is ripe for it.'[66]

The exchange of letters between Sören Gunnarsson and Hans Palmstierna is a single example. It is not possible to draw any far-reaching conclusions about how people in general were affected by, and involved in, the environmental debate on the basis of one such case. Nonetheless, the correspondence does indicate that knowledge was circulating in society. The social distance between a layperson and a scientist was not so great that it could prevent the establishment of a dialogue characterized by trust. Gunnarsson took his own initiatives; he sent articles and asked questions, and Palmstierna answered at length. The exchange of letters shows a genuine desire on both sides to create change and channel

---

64 Letter from Hans Palmstierna to Sören Gunnarsson, 6 November 1967, 452/3/2 (HP ARBARK).
65 Letter from Hans Palmstierna to Sören Gunnarsson, 28 November 1967, 452/3/2 (HP ARBARK). For a study of the Young Philosophers, see Alexander Ekelund, *Kampen om vetenskapen: Politik och vetenskaplig formering under den svenska vänsterradikaliseringens era* (Gothenburg: Daidalos, 2017).
66 Letter from Hans Palmstierna to Sören Gunnarsson, 28 November 1967, 452/3/2 (HP ARBARK).

commitment. It also indicates how the major perspectives regarding survival were hooked into local issues. In addition, it appears from the letters that both men felt that something had happened recently. A social breakthrough in knowledge had occurred.

A statement of this kind, however, obviously raises many questions. What was the situation like before? What had led to the breakthrough? And did knowledge and expertise really circulate in completely different ways in the autumn of 1967 than, say, in 1955 or 1965? Addressing this type of question requires a different approach than the empirical in-depth study of a limited period. In the next chapter, I will therefore change lenses and take a broader look at the first decades of the postwar period. I will supply an overall characterization of vital lines of development from the late 1940s to the summer of 1967, the period in which knowledge about a global environmental crisis was formed.

3
The route to the breakthrough, 1948–1967

In the autumn of 1967, scientific warnings of an impending global catastrophe were nothing new. Knowledge that humanity posed a threat to its own survival had been circulating throughout the postwar period. At first the focus lay on the dramatic threat of a nuclear war causing total annihilation. In parallel with this, equally serious discussions began about overpopulation and dwindling natural resources. Knowledge about a global environmental crisis emerged in, and was shaped by, this broader historical context.

International environmental history research has highlighted the late 1940s as a particularly significant era. That was when a new understanding was established of how humanity, nature, the world, and the future were connected. The very concept of 'the environment' gained a new meaning. Previously the term had referred to the external circumstances which affected humanity. Now it began to be used in order to indicate how human action was reshaping the world. Humanity was regarded as a force of nature and a danger to itself.

Paul Warde, Libby Robin, and Sverker Sörlin stress that this new understanding of the environment was primarily integrative. It assembled a range of problems, challenges, and ideas into a new and more complex whole. This gave rise to a 'modern catalogue of environmental problems', including overpopulation, erosion, industrial waste, overfishing, and water scarcity. The problems themselves were far from unknown, but the overall scientific approach was new. By viewing the various phenomena as aspects of one and the same global complex of problems, the individual problems also came to be regarded as survival issues.[1]

---

1 Ann-Mari Sellerberg, *Miljöns sociala dynamik: Om ambivalens, skepsis, utpekanden, avslöjanden m.m.* (Lund: Department of Sociology, Lund University, 1994); Linnér, *The World Household*, pp. 127–151; Hamblin,

Two influential works from this period are Fairfield Osborn's *Our Plundered Planet* and William Vogt's *Road to Survival*. Both were published in the United States in 1948 and became international bestsellers. In the spring of 1949, the Swedish translation of Osborn's work was marketed as an 'unusual book about a terrible threat to humanity'. The ad asked whether humanity was 'undermining the foundations of its civilization?'.[2] In Vogt's book, the author argued that '[b]y excessive breeding and abuse of the land mankind had backed itself into an ecological trap'. Drastic measures were necessary to avoid a global collapse, and we must all reorientate ourselves in our relationship to the world we live in. 'We can no longer believe valid our assumption that we live in independence' but must instead thoroughly learn 'our dependence upon the earth and the riches with which it sustains us'.[3]

The American warning voices immediately acquired an interpreter in Sweden: Georg Borgström. On the radio in December 1948, he warned of devastating global famines. The following year he wrote, in the preface to the Swedish edition of *Our Plundered Planet*, that 'humanity's traces in nature are terrifying'.[4] Through debate articles in the press, new radio lectures, and such books as *Jorden – vårt öde* [The earth – our destiny] (1953) and *Mat för miljarder* [Food for billions] (1962), Borgström continued to forcefully deliver his message.[5] In the autumn of 1967, he was one of the researchers who contributed to *Människans villkor*.

Another early voice of warning in Sweden was the author, botanist, and member of the Swedish Academy Sten Selander. He was chairman of the Swedish Society for Nature Conservation and wrote the extensive work *Det levande landskapet i Sverige* [The living landscape in Sweden] (1955). It combined an older nature-conservation tradition with the new concept of the environment. Selander also regularly wrote contributions to *Svenska Dagbladet's* prestigious daily essay section 'Under strecket'. There he stated that, over the past century,

---

Arming Mother Nature; Selcer, *The Postwar Origins*; Warde, Robin, and Sörlin, *The Environment*, pp. 1–24.

2 Anon., 'Hotet mot jorden', *DN*, 20 April 1949.
3 William Vogt, *Road to Survival* (New York: W. Sloane Associates, 1948), pp. 284–286.
4 Georg Borgström, 'Preface/Förord' in Fairfield Osborn, *Vår plundrade planet* (Stockholm: Natur & kultur, 1949), p. 8.
5 Linnér, *The World Household*; Engh, 'Georg Borgström and the Population–Food Dilemma'.

humanity had gone from being an animal species to being a force of nature. Humanity had 'launched a whole new geological epoch, the human-ruled one'.[6]

Borgström, Selander, Vogt, and Osborn did not operate in obscurity. They claimed a place on the public stage and attracted considerable attention. A search in their books can easily discover comments that could have been uttered by Hans Palmstierna in the autumn of 1967. Even so, it must be stressed that they were not perceived as environmental debaters in their own time. Not even the leading actors were explicitly aware of the integrated understanding emphasized by Warde, Robin, and Sörlin. As was pointed out above, in the Swedish public debate up until the middle of the 1960s 'the environment' referred to something more limited. In November 1962 when *Dagens Nyheter* wrote that 'the environmental debate had taken off', it was referring to discussions about street and urban planning in Stockholm – not to any global set of problems.[7]

There were, however, significant contexts in the 1940s and 1950s where the new environmental understanding was more prominent. One such context highlighted by Warde, Robin, and Sörlin was the conference *Man's Role in Changing the Face of the Earth*, which was held at Princeton in June 1955. It brought together seventy-three researchers from all over the world to discuss the global challenges facing humanity. By assembling experts from various fields – mainly natural scientists, but also a small number of social scientists and humanities scholars – the conference laid a foundation for applying a broader scientific approach to the environmental field. The importance of expanded international cooperation was also accentuated by major ventures such as the International Geophysical Year from 1957 to 1958. This type of large-scale scientific collaboration was crucial to the formation of the concept of the environment and the new integrated understanding of it. An aggregation of expertise occurred, and around 1960 the term 'environmental sciences' was coined.[8] What, then, characterized this new understanding?

---

6 Sten Selander, 'Historia vid landsvägskanten', *SvD*, 3 July 1955. For a discussion of the 'Under strecket' essays' significance in the history of the Swedish press, see Johan Östling, 'En kunskapsarena och dess aktörer: Under strecket och kunskapscirkulation i 1960-talets offentlighet', *Historisk tidskrift* 140.1 (2020).

7 Anon., 'Fart på miljödebatten', *DN*, 8 November 1962.

8 Warde, Robin, and Sörlin, *The Environment*; Warde and Sörlin, 'Expertise for the Future', pp. 47–49.

The new environmental concept of the postwar era

The new environmental concept was built around four dimensions: future, expertise, trust in numbers, and scale and scalability. Together they shaped and made possible a qualitatively new understanding of the human situation. The idea that there existed a global environment which humans influenced, and also scientific knowledge about a global environmental crisis, emerged symbiotically. The histories of ideas and science intersect with each other here, but also with political history. The institutions and cooperative bodies which were constructed during the postwar period, especially within and via the UN, were of decisive importance to the emergence of the new understanding.[9] But what did these four dimensions involve?

'Future' refers to the temporal direction of the new environmental concept and the threat of a global catastrophe. People began to believe that it was possible to gain knowledge about what was to happen via a scientific route – in broad outline if not precisely. Scientific forecasts and future scenarios thus came to play a central role. This distinguished the concept of the environment from that of nature. The latter had a temporal direction towards the past, and it contained a streak of nostalgia about an original condition which had been lost and could possibly be restored. Conversely, 'environment' referred to a crisis in the making.[10]

Scientific interest in the future grew markedly during the first few decades of the postwar period. Many actors and institutions were involved in this development. Major ventures were made – not least in the United States – within academia, the military, and industry. Warde, Robin, and Sörlin talk about this development as 'a futurological soup'. In that soup, the new environmental concept became an important ingredient.[11] Recent years have seen extensive empirical studies of how the interdisciplinary field of future studies emerged. One important insight conveyed by them is that futures research

---

9 Warde, Robin, and Sörlin, *The Environment*, pp. 1–2, 14; Selcer, *The Postwar Origins*; Simone Schleper, *Planning for the Planet: Environmental Expertise and the International Union for Conservation of Nature and Natural Resources, 1960–1980* (New York: Berghahn Books, 2019).
10 Warde, Robin, and Sörlin, *The Environment*, pp. 14–15, Warde and Sörlin, 'Expertise for the Future', p. 39.
11 Warde, Robin, and Sörlin, *The Environment*, p. 15.

in the 1950s and 1960s was not one single thing but rather several contradictory ones. The future was an elastic, and highly ideological, concept.[12]

'Expertise' refers to the growing scientific legitimacy with which certain actors and institutions spoke about the future. This expertise was integrative rather than specialized and empirical. Nevertheless, it was important for those who positioned themselves as experts on the future that they possessed some form of specialist expertise. Such expertise guaranteed their status as scientists, which was necessary if they were to speak with authority about key issues for humanity.[13]

Hans Palmstierna is a typical example of this group. His legitimacy as a scientist was based on the fact that he was an associate professor of chemistry. The main theme of his scientific warnings, however, was overpopulation. In this respect he made himself a spokesman for science as a whole rather than for his own field of research. This pattern recurs in all the scientific warning voices during the postwar period. The new environmental concept was so comprehensive that no single person had more than fragmentary expert knowledge. Therefore the expertise had to be aggregated, either by an actor turning himself into a spokesperson, such as Hans Palmstierna, or by a number of people joining forces, such as the twelve authors of *Människans villkor*.

'Trust in numbers' refers to the way in which scientific knowledge was aggregated in practice. This was done through the quantification

---

12 Jenny Andersson, 'The Great Future Debate and the Struggle for the World', *American Historical Review* 117.5 (2012); Elke Seefried, 'Steering the Future: The Emergence of "Western" Futures Research and its Production of Expertise, 1950s to the Early 1970s', *European Journal of Futures Research* 29.2 (2014); Andersson and Rindzevičiūtė (eds), *The Struggle for the Long-term*; Elke Seefried, 'Reconfiguring the Future? Politics and Time from the 1960s to the 1980s', *Journal of Modern European History* 13.3 (2015); Elke Seefried, *Zukünfte: Aufstieg und Krise der Zukunftforschung* (Berlin: De Gruyter Oldebourg, 2015); David Larsson Heidenblad, 'Tillbaka till framtiden', *Statsvetenskaplig tidskrift* 118.2 (2016); Gustav Holmberg, 'Framtiden: Historikerna blickar framåt', in Gunnar Broberg and David Dunér (eds), *Beredd till bådadera: Lunds universitet och omvärlden* (Lund: Lund University, 2017); Jenny Andersson, *The Future of the World: Futurology, Futurists, and the Struggle for the Post-Cold War Imagination* (New York: Oxford University Press, 2018).

13 Warde, Robin, and Sörlin, *The Environment*, pp. 15–16; Warde and Sörlin, 'Expertise for the Future', pp. 41, 48–50.

of data. The changes that were being caused by humans were visibly presented in the form of diagrams and steeply rising curves. With the help of numbers, scientifically based predictions could be made. Lines of development could be extrapolated and the significance of different variables discussed. Expertise and knowledge about the future rested on a conviction that measurements and calculations comprised an objective and neutral knowledge base.

One crucial factor which allowed numbers to acquire this status was that they were able to illustrate change over time. It was fundamental to the new understanding of the environment that such a change was occurring, and that humans were the cause. Quantifications legitimized this view, functioning as arguments for political action. In addition, numbers could be transferred between disciplines and used to construct models. In this respect computers became increasingly important. They enabled the making of ever-larger calculations and the simulation of future developments.[14]

'Scale and scalability' means that the new environmental concept connected local phenomena with global conditions. The environmental concept was useful on all levels. Environmental impact might refer to how an individual industrial plant was polluting a specific watercourse, but it could also refer to the impact of the burning of fossil fuels on the Earth's climate. The latter is an example of a global process which came to be linked to local phenomena such as storms and forest fires.

That the same methods and techniques could be used to achieve knowledge at various levels was also a significant factor. Numbers, models, and simulations could be done on all scales – from individual ecosystems to the entire planet's biosphere. The quantitative approach made it possible to combine local data in order to comment on wider contexts. Scale and scalability reinforced trust in scientists who adopted a comprehensive approach to the entire set of environmental problems. Their specialist expertise about a phenomenon

---

14 Paul N. Edwards, *A Vast Machine: Computer Models, Climate Data, and the Politics of Global Warming* (Cambridge, MA: MIT Press, 2010); Howe, *Behind the Curve*; Warde and Sörlin, 'Expertise for the Future', pp. 44, 49–51; Warde, Robin, and Sörlin, *The Environment*, pp. 16–17; Theodore Porter, *Trust in Numbers: The Pursuit of Objectivity in Science and Public Life* (Princeton, IL: Princeton University Press, 1995). See also Johan Fredrikzon, *Kretslopp av data: Miljö, befolkning, förvaltning och den tidiga digitaliseringens kulturtekniker* (Lund: Mediehistoriskt arkiv, 2021).

on a specific scale also gave them the authority to comment on problems that were on completely different scales: they established a scientific meta-expertise.[15]

The four dimensions of the new environmental concept were closely linked. They were one another's prerequisites, and they reinforced one another. Scalable numbers made it possible for scientific expertise to comment on a threatening future. However, another historical context existed which was crucial to the formation of the new environmental concept: the threat of a devastating nuclear war.

The nuclear weapons threat

The bombing of Hiroshima and Nagasaki ended the Second World War and launched a new era. At the time, it was said that humanity had entered the atomic age. That age came to be characterized by the realization that humanity now possessed the power to destroy itself. A full-scale nuclear war could kill unimaginable numbers of people within a short time. In the long term, radioactive fallout would make the Earth uninhabitable. Knowledge about this new situation began circulating immediately after the bombings.[16]

In recent years, historians have asserted that the nuclear-weapons threat made new ways of thinking about global threats possible. The nuclear threat made people more sensitive to perceptions of planetary contexts and deadly risks. The threat of human-caused planetary destruction gradually expanded from nuclear war to overpopulation and insufficient natural resources. From there it was a short step to the global environmental crisis. All of these threats were explained and discussed using similar words and images, unified by an apocalyptic use of language. Scientific warning voices played a key role.[17]

---

15 Warde and Sörlin, 'Expertise for the Future', p. 42; Warde, Robin, and Sörlin, *The Environment*, pp. 17–18.
16 Paul S. Boyer, *By the Bomb's Early Light: American Thought and Culture at the Dawn of the Atomic Age* (Chapel Hill, NC: University of North Carolina Press 1985/1994), pp. 5, 12, 22.
17 M. Jimmie Killingsworth and Jacqueline S. Palmer, 'Millennial Ecology: The Apocalyptic Narrative from "Silent Spring" to "Global Warming"', in Carl G. Herndl and Stuart C. Brown (eds), *Green Culture: Environmental Rhetoric in Contemporary America* (Madison, WI: University of Wisconsin Press, 1996); Holger Nehring, 'Cold War, Apocalypse, and Peaceful Atoms: Interpretations of Nuclear Energy in the British and West German Anti-Nuclear Weapons Movements, 1955–1964', *Historical Social Research/Historische*

The first joint initiative was taken in March 1946 with the publication of the report *One World or None*, in which the world's leading nuclear physicists sought to explain the full implications of the atomic bomb. The report's authors included Albert Einstein and Robert Oppenheimer, and the preface was written by the Danish physicist Niels Bohr. The scientists tried to awaken the general public to a realization of the life-threatening situation. The solution they envisioned was expanded internationalism. Time was short, and humanity's survival was at stake. However, the political realities of the Cold War as well as nascent anti-communist sentiments in the United States would soon silence the scientific alarms.[18]

A second phase of scientific activism emerged in the mid-1950s and was sparked by the American nuclear test explosions on the Bikini Atoll in 1954. Their radioactive fallout had an adverse effect on Japanese fishermen, who were far outside the specified safety zone. People's fear of radioactivity grew, and resistance to atmospheric nuclear testing emerged. The philosopher Bertrand Russell and the scientist Linus Pauling assumed leading roles, launching petitions and collecting signatures. An international scientific cooperation organization, Pugwash, was also created in order to promote peace and disarmament.[19]

The scientists' actions occurred in parallel with the emergence of new social protest movements. Natural scientists often played key roles in peace and disarmament campaigns. Some historians believe that these movements were the direct predecessors of the environmental movements of the 1970s.[20] In the case of Sweden, these issues were

---

Sozialforschung 29.3 (2004); Egan, *Barry Commoner and the Science of Survival*; Hamblin, *Arming Mother Nature*; Joseph Masco, 'Bad Weather: The Time of Planetary Crisis', in Martin Holbraad and Morten Axel Pedersen (eds), *Times of Security: Ethnographies of Fear, Protest, and the Future* (New York: Routledge, 2013); Warde and Sörlin, 'Expertise for the Future'; Frank Uekötter (ed.), *Exploring Apocalyptica: Coming to Terms with Environmental Alarmism* (Pittsburgh, PA: Pittsburgh University Press, 2018); Toshihiro Higuchi, *Nuclear Weapons Testing and the Making of a Global Environmental Crisis* (Stanford, CA: Stanford University Press, 2020).

18 Boyer, *By the Bomb's Early Light*, pp. 76–81; Jessica Wang, 'Scientists and the Problem of the Public in Cold War America 1945–1960', *Osiris* 17.1 (2002).

19 Alison Kraft, Holger Nehring, and Carola Sachse, 'The Pugwash Conference and the Global Cold War: Scientists, Transnational Networks, and the Complexity of Nuclear Histories', *Journal of Cold War Studies* 20.1 (2018).

20 Nehring, 'Cold War, Apocalypse, and Peaceful Atoms'; Nehring, 'Genealogies of the Ecological Moment'.

brought to the fore in the late 1950s as a result of plans to develop Swedish nuclear weapons. An intense debate flared up. The resistance was organized via Aktionsgruppen mot svenskt atomvapen [Action Group against Swedish Nuclear Weapons].[21]

The Swedish plans to become a nuclear power were not realized, and the debate and opposition subsided. However, concern about nuclear war remained strong in the society of the early 1960s. Sweden's civil defence system was one of the most developed and best funded in the world.[22] The Cuban Missile Crisis of 1962 marked a culmination: never before had nuclear war been so close to becoming a reality.[23] The following year, however, the United States, the Soviet Union, and the United Kingdom signed the Partial Nuclear-Test-Ban Treaty, which banned atmospheric tests. A period of detente between the Great Powers began. For almost two decades, the threat of nuclear war was relatively absent from the public debate and cultural life. Paul Boyer has labelled the period from 1963 to the early 1980s 'the Big Sleep'.[24]

---

21 Anna-Greta Nilsson Hoadley, *Atomvapnet som partiproblem: Sveriges socialdemokratiska kvinnoförbund och frågan om svenskt atomvapen 1955–1960* (Stockholm: Almqvist & Wiksell, 1989); Wilhelm Agrell, *Svenska förintelsevapen: Utvecklingen av kemiska och nukleära stridsmedel 1928–1970* (Lund: Historiska media, 2002); Karl Haikola, 'Atombomben och det moderna samhället: Om framstegstankens roll i motståndet mot svensk atombomb 1956–1961' (Bachelor's degree research essay, Department of History, Lund University, 2014); Thomas Jonter, *The Key to Nuclear Restraint: The Swedish Plans to Acquire Nuclear Weapons* (London: Palgrave, 2016).

22 For studies of Swedish civil defence culture see Marie Cronqvist, 'Bilder från nollpunkten: Visualiseringar av atomålderns urbana apokalyps', in Eva Österberg and Marie Lindstedt Cronberg (eds), *Våld: Representation och verklighet* (Lund: Nordic Academic Press, 2006); Marie Cronqvist, 'Utrymning i folkhemmet: Kalla kriget, välfärdsidyllen och den svenska civilförsvarskulturen 1961', *Historisk tidskrift* 128.3 (2008); Marie Cronqvist, 'Det befästa folkhemmet: Kallt krig och varm välfärd i svensk civilförsvarskultur', in Magnus Jerneck (ed.), *Fred i realpolitikens skugga* (Lund: Studentlitteratur, 2009); Marie Cronqvist, 'Survival in the Welfare Cocoon: The Culture of Civil Defense in Cold War Sweden', in Annette Vowinckel, Marcus Payk, and Thomas Lindenberger (eds), *Cold War Cultures: Perspectives on Eastern and Western European Societies* (New York: Berghahn Books, 2012).

23 Alice L. George, *Awaiting Armageddon: How Americans Faced the Cuban Missile Crisis* (Chapel Hill, NC: University of North Carolina Press, 2003); David Larsson Heidenblad, *Vårt eget fel: Moralisk kausalitet som tankefigur från 00-talets klimatlarm till förmoderna syndastraffsföreställningar* (Höör: Agerings, 2012), pp. 171–207.

24 Boyer, *By the Bomb's Early Light*, p. 355.

That is a reasonable description of the nuclear-weapons threat as an isolated global menace. However, if it is regarded as being one of several interconnected hazards, a different picture emerges. Then the 1960s do not look like a sleeping decade, but rather like a period when anxiety broadened and the focus shifted from nuclear war and radioactivity to overpopulation and environmental destruction. Many cultural connections were made between these threats, not least the growing concern about 'the population explosion'. For example, Georg Borgström illustrated the demographic trend with a diagram in the form of a mushroom cloud, and Paul Ehrlich's international bestseller was entitled *The Population Bomb* (1968).

In the dawning environmental debate of the 1960s, radioactivity and its link to cancer played a similar role. This invisible threat came to influence society's perception of other environmentally hazardous substances, not least chemical pesticides such as DDT. The starting point for this debate was Rachel Carson's book *Silent Spring* (1962). Carson warned that birdsong would be silenced and that human DNA risked being broken down. She wrote that such 'poisons' should not be called 'insecticides' but 'biocides'.[25] Her book caused a huge debate in the United States. Sharp lines of division were created between industrial interests and nature conservationists. Scientists spoke out on both sides, but politicians and lay people were also involved. The environmental debate entered a new and increasingly intense phase. Carson herself, however, did not experience much of it. In April 1964 she died of breast cancer.[26]

The early 1960s debates in Sweden

The American debate over biocides did not pass unnoticed in Sweden. *Silent Spring* was translated more or less immediately and was published in March 1963 as *Tyst vår*. In Carson's book, the new concept of the environment plays a key role. She asserted that along with 'the possibility of the extinction of mankind by nuclear war', the key problem was 'the contamination of man's total environment with such substances of incredible potential for harm'.[27] Her starting point was that over the past twenty-five years, mankind had become a force of nature. Speaking about 'man's assaults upon

---

25 Rachel Carson, *Silent Spring* (Boston: Houghton Mifflin, 1962), p. 8.
26 Lear, *Rachel Carson*; Kroll, 'The "Silent Springs" of Rachel Carson'.
27 Carson, *Silent Spring*, p. 8.

the environment', she added that a 'universal contamination of the environment' was occurring with disturbing rapidity.[28] 'Future generations', she wrote, 'are unlikely to condone our lack of prudent concern for the integrity of the natural world that supports all life.'[29]

The Swedish reception of *Tyst vår* has been studied by Anna Tunlid. She points out that Swedes began discussing the book even before it had been translated. As early as February 1963, the Swedish Natural Science Research Council held a conference on the use of biocides.[30] Simultaneously, zoologist Erik Dahl reviewed the English edition in *Dagens Nyheter*. He stressed the great uncertainty which existed regarding long-term consequences. There were not yet enough 'definite and concrete facts'. However, the possibility of 'increased cancer rates' and 'hereditary changes' was a cause for concern. Dahl underlined that biologists and physicians were uneasy about developments. He hoped that Carson's book would become 'a kindling spark in a public debate that we have been lacking for far too long'.[31] That hope would be fulfilled.

The Swedish biocide debate was particularly intense during the spring and summer of 1963. Just as in the United States, opinion was divided. On one side stood representatives of the chemical industry and authorities such as Statens växtskyddsanstalt [the Swedish Plant Protection Agency] and Giftnämnden [the Poisons Board]. These voices argued that Carson's portrayal was one-sided and tendentious. On the opposite side, proponents of nature conservation maintained that *Tyst vår* was factual and moderate. Overall, the debate focused very much on biocides; there was no in-depth criticism of modern industrial society. Both sides were profoundly confident that scientific expertise and technological solutions could solve the problem.[32]

Most of the debaters were scientists. They wrote on editorial and cultural pages in the general press, but they also developed their reasoning in learned and specialist journals. The biocide debaters primarily wrote to and for each other and did not seek to mobilize

---

28 *Ibid.*, p. 6.
29 *Ibid.*, p. 13.
30 Anna Tunlid, 'Människan och naturens överlevnad: Mottagandet av *Tyst vår* och *Plundring, svält, förgiftning* i den svenska miljödebatten' (Bachelor's degree research essay, Department of Philosophy, Lund University, 1994), p. 15.
31 Erik Dahl, 'När fågelsången tystnar', *DN*, 8 February 1963.
32 Tunlid, 'Människan och naturens överlevnad', pp. 15–22.

the general public. They broadly agreed that more research was needed in this area and that large-scale studies of the effects of environmental toxins must be initiated. The debaters focused most of their attention on how animal and birdlife were being affected. It was nature that was under threat, not humanity.[33]

The Swedish biocide debate became a catalyst for new research projects. Particular focus was placed on mercury-treated seeds and their impact on birdlife. The National Veterinary Institute and the Swedish Ornithological Society played important roles in this respect.[34] However, the biocide debate also coincided with the Swedish government assuming an increasingly active nature-conservation role during the early 1960s.[35] Official reports were commissioned, and special committees were set up. The previously mentioned *Forskningsberedningen* was crucial in this context. It ensured that the step from scientific debate to discussions at the highest political level was short. The appointment of the 1964 government enquiry into natural resources, which went on to present its report in the autumn of 1967, was one result of this close connection.

The debates, official enquiries, and research initiatives of the early 1960s were intertwined. Science and politics were communicating vessels. The new environmental concept was also sporadically in evidence, for example in *Tyst vår*. Despite this, there was no general talk about an environmental debate or environmental debaters. It was nature and nature conservation that were on the agenda. From 1965 onwards, though, there are indications of a linguistic reorientation. It is, for instance, evident in the Riksdag debates, which began using the concept of 'environmental protection' that year.[36] There are also traces in the press. Bengt Lundholm, secretary of the then-ongoing enquiry into natural resources, said in an interview that it was 'characteristic of the new approach to the problems that the term "nature conservation" is increasingly being replaced by the term "human protection"'.[37] An unsigned editorial in *Dagens Nyheter* further proposed that the term

---

33 *Ibid.*, pp. 15–22.
34 *Ibid.*, pp. 20–22.
35 Erland Mårald and Christer Nordlund, 'Modern Nature for a Modern Nature: An Intellectual History of Environmental Dissonances in the Swedish Welfare State', *Environment and History* 26.4 (2020).
36 *Register till Riksdagens protokoll med bihang 1961–1970. Bd 3, Sakregister, L–Ö* (Stockholm: Riksdagen, 1893–1971).
37 Anon., 'Amerikansk kontroll avslöjar DDT i luften', *DN*, 16 February 1965.

'environmental protection' should replace the 'somewhat worn concept' of 'nature conservation'. The latter brought to mind 'the aesthetic requirement not to throw away tins in the woods' rather than 'the vital need to curb the poisoning and imbalance of our living environment and our natural resources that are a consequence of civilization'.[38] This comment is a foretaste of the direction that the debate was about to take. Its tone of voice was about to be raised to a higher pitch.

Unlike the biocide debate of the early 1960s, the upcoming Swedish environmental debate came to focus on humanity's survival. Even so, the actual substantive issues did not disappear from the agenda. On the contrary; as the decade continued, mercury and other environmental toxins received even more attention. This was done by linking the biocides with, and integrating them into, a larger global complex of problems. The following section presents a particularly illustrative example of this process.

Rolf Edberg and the three global threats

In the autumn of 1966, Rolf Edberg, Sweden's then ambassador in Oslo, published the book *Spillran av ett moln* [published in English as *On the Shred of a Cloud* (University of Alabama Press, 1969)]. The book has long been central to Swedish environmental-history research, and it has been described as something of a breakthrough in the general public's awareness of an environmental perspective.[39] Published simultaneously in the three Scandinavian languages, *Spillran av ett moln* immediately made an impression in their respective national public spheres. In retrospect, it has become regarded as an early document of environmental awakening. The global environmental crisis was only one aspect of Edberg's message, though. The theme throughout the book was that modern industrial civilization posed a deadly threat to life on Earth in a variety of ways. Edberg addressed the issues of nuclear war and

---

38 Anon., 'Miljöskyddet', *DN*, 27 March 1965.
39 Anshelm, *Socialdemokraterna och miljöfrågan*, p. 14; Eva Friman, 'Domedagsprofeter och tillväxtpredikanter – debatten om ekonomisk tillväxt och miljö i Sverige 1960–1980', *Historisk tidskrift* 121.1 (2001), p. 33; Björn-Ola Linnér, *Att lära för överlevnad: Utbildningsprogrammen och miljöfrågorna 1962–2002* (Lund: Arkiv, 2005), p. 65. See also Berntsen, *Grønne linjer*, pp. 179–180.

accelerating population growth both prior to and in more detail than the issue of environmental destruction.[40]

Spillran av ett moln is difficult to categorize in terms of genre. It has some of the characteristics of a popular-science debate book, but it is equally a pastoral travelogue. Besides, it is a philosophical contemplation of existence. Edberg's literary ambitions are obvious. Poetic formulations overshadow numbers and scientific facts. This distinguishes his book from *Plundring, svält, förgiftning* and *Människans villkor*. But then, Rolf Edberg was not a scientist. In fact, he had no academic credentials to speak of. Immediately after graduating from secondary school, he began working for the social democratic press and simultaneously launched a political career. During the 1940s and 1950s, he represented the Social Democrats in the Riksdag for two mandate periods. Edberg was an unusual voice in the emerging environmental debate. He was the learned layman who used literary techniques in order to turn himself into a spokesman for science.

The overall narrative structure of *Spillran av ett moln* consisted of a hiking trip in the Norwegian mountains made by Edberg together with his 17-year-old son – 'the travelling companion'. Their co-existence was wordless, but it added topicality and concreteness to the global state of emergency. In his account, Edberg combined his concern for his son's future with lyrical depictions of nature. He depicted the long history of life on Earth, emphasizing that humanity was merely a small part of a large web. Edberg described the thin film covering the parts of the Earth where human life is possible as 'a marginal home for marginal beings', surrounded by 'the cosmic realms of the dead'. The whole history of the human race had taken place in that sphere, but the existence of humanity had changed utterly at a single stroke: 'It became a different one the moment humans acquired opportunities to exterminate their own species.'[41]

The nuclear threat was the book's starting point and vital thread. In Edberg's words, his generation had crossed the boundary of an epoch. The destructive power conjured up by humans 'lies beyond the

---

40 David Larsson Heidenblad, 'Ett ekologiskt genombrott? Rolf Edbergs bok och det globala krismedvetandet i Skandinavien 1966', *Historisk tidskrift* (NO) 95.2 (2016).
41 Rolf Edberg, *Spillran av ett moln: Anteckningar i färdaboken* (Stockholm: Norstedts, 1966). The following page references are to the collective edition *På Jordens villkor: En trilogi om människan och hennes värld* (Stockholm: Norstedts, 1974), pp. 14–16.

concrete capacity to imagine'. It extended not only in the present but also 'deep down into unborn generations'. The contrast to the Norwegian mountain expanses was sharp and hard to grasp. 'The chattering brook by the hiking trail and the threat of global poisoning – something here doesn't add up. Yet both are realities.'[42] It seemed even more absurd to him that nuclear war could be triggered by accident.[43]

Edberg's view of humanity was characterized by a biological, evolutionary, and historical point of view. Humans were what they had always been and acted as they had always done. This was cause for deep concern. 'The sum of human history', he wrote, 'becomes the story of one species, more self-destructive than any other in Creation.'[44] What was needed was to lead humanity's ancient instincts and habitual behaviours down new paths. Otherwise the collective suicide would soon be complete. To avoid this, Edberg argued, humanity must fundamentally change its way of thinking and being. He believed this could happen if human beings realized their cosmic insignificance and began to see themselves as part of a larger whole.[45]

From the threat of nuclear war and the cosmic expanses, Edberg shifted his focus to the population explosion. He pointed out that all the lines on charts were rising at a ferocious speed; they 'are bolting such that all numbers are quickly becoming out of date'. The reason was that humankind, by means of science and technological development, had 'freed itself from the brakes that hold other species back'. The future was bleak despite the fact that all the necessary knowledge existed. 'International bodies and the mass media are showering us with facts about the population explosion', yet it was continuing unabated. If nuclear war did not become a reality, 'overpopulation could become an even more tangible threat to a future generation'.[46] Edberg envisioned a world that was on the brink of catastrophic famines and experiencing constant crises, conflicts, and global anarchy. In this situation, the global threats were united into one. The population explosion risked 'ultimately becoming the direct trigger of the nuclear bomb'.[47]

Edberg, however, also perceived a third threat to the survival of the human race: the global environmental crisis. In this context,

---

42 *Ibid.*, pp. 14, 18–21.
43 *Ibid.*, p. 23.
44 *Ibid.*, p. 48.
45 *Ibid.*, p. 52.
46 *Ibid.*, pp. 107–112.
47 *Ibid.*, pp. 117–118.

history – not least the decline and fall of classical civilizations – played a central role in his reasoning.[48] 'The most imposing ruins along humanity's path are not found at the Acropolis or the Roman Forum', he wrote, but 'are encountered in ruined landscapes.' This was a classic tragedy which happened again and again. In their short-term arrogance, humans subjugated nature and thereby destroyed the long-term basis of their own existence. Edberg perceived this pattern recurring in all countries with a long cultural history. The fate of the ancient world 'is becoming that of the globe'. For him, this was an 'indirect form of cannibalism, more macabre than the direct one ever was, because it affects generations as yet unborn'.[49]

To Edberg, overpopulation and the destruction of nature went hand in hand. He described humanity as 'the skin cancer of the Earth' which 'has etched deep wounds and scratches on the face of the Earth'. He singled out chemical pesticides as 'something new which has been added to the old story of humanity's reckless advance on the globe'. Their use was evidence of insensitivity to the context and interdependence of all living things. The long-term consequences were unpredictable, but all indications were that they would be devastating. Humankind was in a stage of 'accelerated pollution of our entire environment', he wrote, a 'cancer crisis of the globe'.[50]

To Edberg, it was perfectly clear that humanity had placed itself in this dire situation. It alone was responsible for the 'triple threat' it faced. But 'there will be no Ragnarök unless humans themselves initiate it', he emphasized. The threats could be averted if 'our concepts and our actions are radically adapted to the fundamentally altered conditions we ourselves have created'. He placed his hope in some form of world government and the establishment of a new world view. Humanity could no longer afford to put itself and its immediate needs first. If it did, that would be the end of the human race.[51]

---

48 Isak Hammar, 'Det ständiga fallet: Romarriket som politisk resurs i samtiden', *Statsvetenskaplig tidskrift* 117.3 (2015); David Larsson Heidenblad and Isak Hammar, 'A Classical Tragedy in the Making: Rolf Edberg's Use of Antiquity and the Emergence of Environmentalism in Scandinavia', *International Journal of the Classical Tradition* 24.2 (2017); Gustaf Johansson, *När man skär i nuet faller framtiden ut: Den globala krisens bildvärld i Sverige under 1970-talet* (Uppsala: Uppsala University, 2018), pp. 95–110.
49 Edberg, *Spillran av ett moln*, pp. 127–131, 135.
50 Ibid., pp. 151–165.
51 Ibid., pp. 169–187.

Spillran av ett moln shows in an almost overly explicit way how the postwar environmental consciousness was shaped by the historical context in which it arose – a time and a culture deeply marked by the Cold War, the nuclear arms threat, and realizations about the sufferings of the world's poor. But what impact did Edberg's book actually have in 1966? Did it really mark the breakthrough of an ecological point of view in the public consciousness? My studies of the book's contemporaneous Scandinavian reception do not confirm this to be so. The extensive extant review material shows that the book was perceived as topical and successful, but hardly as pioneering. This might have been due to the very fact that Edberg adopted a holistic approach to humanity's crucial issues. Nuclear war and overpopulation had been discussed for a long time. Given the prominent place these themes have in the book, it is not surprising – nor in itself incorrect – that his contemporaries felt he was conveying well-known warnings.[52]

Rolf Edberg's fame as an environmental pioneer emerged later. Within a couple of years, his book came to be read in new ways and appeared in new editions. The same was true of Rachel Carson. *Silent Spring* circulated in a different way in 1970 than it had done back in the early 1960s. By 1970, it had become a cornerstone in a larger context in which a chorus of scientific voices warned of a global environmental crisis. In Sweden, the big change occurred in the autumn of 1967. But the breakthrough did not happen like a bolt from the blue. Two of the key scientific actors, Hans Palmstierna and Karl-Erik Fichtelius, had begun to warn of an impending catastrophe as early as 1966. The knowledge arena they used was *Dagens Nyheter*'s cultural page.[53] By retracing their footsteps we can see how they, and their interaction with other actors, paved the way for the great breakthrough.[54]

The survival debaters

The first of the two scientists to raise his voice was Hans Palmstierna, and the time was the spring of 1966. The previous year he had

---

52 Larsson Heidenblad, 'Ett ekologiskt genombrott?', p. 263.
53 Anders Frenander, *Debattens vågor: Om politisk-ideologiska frågor i efterkrigstidens svenska kulturdebatt* (Gothenburg: Department of History of Ideas and Science, University of Gothenburg, 1998), pp. 138–159; Johan Stenfeldt, *Dystopiernas seger: Totalitarism som orienteringspunkt i efterkrigstidens svenska idédebatt* (Höör: Agerings, 2013), pp. 116–117.
54 The following is based on Larsson Heidenblad, 'Överlevnadsdebattörerna'.

begun writing articles for *Dagens Nyheter* on science topics, but those early works were of an explanatory and apolitical nature. In March 1966 he addressed the population question and adopted a more strident tone. In dramatic terms he stated that 'if we do not alter our thinking quickly, inform our fellow human beings of the catastrophe that not only threatens but is already a fact, then we will slide straight into an overpopulated hell of disease and famine within a few decades'.[55] The only solution to the problem that he could see was to restrict the birth rate.

Palmstierna's first article did not attract any attention, but the second one did. 'Malthus och världssvälten' [Malthus and world hunger] was published in May 1966 and started out from the fact that 1966 marked the two-hundredth anniversary of the birth of the British cleric and social philosopher Thomas Robert Malthus. Palmstierna pointed out that Malthus was the first to 'express with mathematical rigour the formula which shows that humanity may steer down into the inferno of overpopulation'. But Palmstierna had little time for his predecessor's appeal to moral restraint. Pious hopes were doomed to fail. 'We do, however, have means in our hands that Malthus did not have', he emphasized, adding that 'we have a huge propaganda apparatus on the radio and television, with and without satellites. We have a reasonably well-functioning press all over the world. This expanded and effective apparatus could be used for propaganda to control the birth rate.'[56]

Shortly afterwards, the editor-in-chief of *Dagens Nyheter*, Olof Lagercrantz, said he had received many compliments about the article.[57] Even more important, though, was that the acclaimed author, engineer, and public intellectual Sven Fagerberg drew attention to it a couple of weeks later. He stated that it was 'every scientist's duty' to comment on the human situation on the basis of his or her own expertise. 'Unfortunately, this happens far too seldom, but there are shining exceptions. As I write this, the most recent example is Hans Palmstierna's article on the population issue.'[58] Fagerberg's appreciative words reached their subject immediately. That same

---

55 Hans Palmstierna, 'Vaccin mot spetälska', *DN*, 17 March 1966.
56 Hans Palmstierna, 'Malthus och världssvälten', *DN*, 3 May 1966.
57 Letter from Olof Lagercrantz to Hans Palmstierna, 11 May 1966, 452/3/2 (HP ARBARK).
58 Sven Fagerberg, 'Målsättning och dubbelmoral', *DN*, 20 May 1966.

day, Palmstierna wrote Fagerberg a letter thanking him for the kind words and inviting him home for dinner.[59]

Their amicable relationship would strengthen. The proposed dinner took place in the summer of 1966. In connection with it, Fagerberg encouraged Palmstierna to write a debate book.[60] The idea took root, and in September Palmstierna told Lagercrantz that he was 'gradually persuaded to try to write a "debate" book about overpopulation, erosion, and the consumer economy'.[61] Shortly thereafter, he published his third major article on the population issue in *Dagens Nyheter*. In it he drew an analogy with the growth curve of a bacterial culture. 'What happens to these creatures can serve as a model for what happens to cultures of organisms in closed systems, e.g. humanity on earth.' For the bacteria in a laboratory flask, an explosive growth phase was followed by an equally rapid logarithmic death phase. 'Humanity, however, differs in one respect from bacteria and mould: it seeks to see into the future and adjust its actions according to the result. Herein lies the opportunity for humanity to control its own growth curve.'[62] The next day, Palmstierna's article was discussed in an editorial. It stated that the information about the factual situation and trends had now undoubtedly begun to make a breakthrough on a broad front. 'Nearly everyone *knows* theoretically today', but 'to *understand* the implications and try to act accordingly' – that was something completely different.[63]

The dire situation weighed heavily on Palmstierna, as is clear from a candid letter to Fagerberg in November 1966. 'I know that I have to write, have to speak', he said. 'You do not know what you started when you urged me to write. I had protected myself behind a wall of supposed inability to write about what concerns us most.' Palmstierna emphasized his own fear of discouragement and depression. He was tormented in and about his new role as a survival debater. 'It is not right that I must be one of those who see that

---

59 Letter from Hans Palmstierna to Sven Fagerberg, 20 May 1966, 452/3/2 (HP ARBARK).
60 Letter from Sven Fagerberg to Hans Palmstierna, 21 July 1966, 452/3/2 (HP ARBARK).
61 Letter from Hans Palmstierna to Olof Lagercrantz, 22 September 1966, 452/3/2 (HP ARBARK).
62 Hans Palmstierna, 'Konsumtion och kritisk befolkningstäthet', *DN*, 13 October 1966.
63 Anon., 'Ständig folkexplosion?', *DN*, 14 October 1966.

things are going to hell, that I must write about it, as if nothing could be done about it.' In the letter he reflected further on the limitations of language. '[I] also detest the term "debate book". As if there were something to debate. Then it is in the same category as the "infidelity debate" and other manias – valuable in a society that believes it will survive, but meaningless in the face of catastrophe.'[64] The deep apprehension and reluctance that Palmstierna expressed shows how the transition process from research to social debate manifested itself for him. The new role was almost a calling, forced on him by scientific clearsightedness. There was no alternative: he had to act.

On Christmas Eve 1966, a new scientist appeared on *Dagens Nyheter*'s culture page to discuss the key issues for humanity. Karl-Erik Fichtelius's insights into the crisis closely resembled those of Hans Palmstierna and Rolf Edberg. 'It is utterly obvious', Fichtelius wrote, 'that a catastrophe of previously unimagined dimensions may befall humanity in the near future. One does not have to be a doomsday prophet to come to that conclusion.' His article focused on the population explosion as well as on weapons of mass destruction, but he was also preoccupied with humanity's biological nature and innate aggression. This was what made humanity so dangerous to itself. 'So something must be done, and done soon, for us to avoid annihilating ourselves.'[65]

Fichtelius openly admitted that he was moving into a new and hard-to-navigate arena. 'It is hazardous for a scientist to discuss political issues', he wrote, because it was so easy to be 'branded a political idiot'. He did not oppose this state of affairs; on the contrary, he emphasized that scientists lacked politicians' expertise in this field. Consequently, he attempted to pass the ball to the politicians, who he felt were the only ones with any prospect of creating peace on earth. 'Politicians are specialists in creating opinion and leading development', he asserted, but they needed to be 'influenced by

---

64 Letter from Hans Palmstierna to Sven Fagerberg, 13 November 1966, 452/3/2 (HP ARBARK). The so-called 'infidelity debate' was carried out in the Swedish culture pages in the mid-1960s and focused on living independently and free from traditional norms. In a Swedish context, the discussion was linked to the emergence of the New Left. For studies of this see Birgitta Jansson, *Trolöshet – En studie i svensk kulturdebatt och skönlitteratur under tidigt 1960-tal* (Uppsala: Uppsala University, 1984).
65 Karl-Erik Fichtelius, 'Om frid på jorden', *DN*, 24 December 1966.

suggestion in order to become emotionally involved in the right way.'[66] Here, perceptive scientists had a key role to play.

Fichtelius's thoughts have clear idealistic overtones. He placed his trust in an enlightened world government which would rule on the basis of scientific information and not be affected by any conflicting goals or interests. His reasoning gravitated towards a global species level. He felt that important answers to humanity's challenges could be found in ethology – the study of animals and their behaviours. Where Palmstierna cited bacterial cultures, Fichtelius focused on jackdaws and dolphins.

The late winter and early spring of 1967 was a period of increasingly intense activity for the two debaters. Both men regularly wrote new articles on many different topics. For Palmstierna, this was an integral part of his work on his book manuscript. His debate articles in *Dagens Nyheter* would become essential parts of *Plundring, svält, förgiftning*.[67] This relationship was discussed with the publisher, who pointed out that a maximum of one-third of a book could have been previously published in a newspaper.[68] However, it was Fichtelius who reflected most extensively on his new role and activities. In March 1967, he said that scientists did manage to make their voices heard from time to time. 'They come with cold numbers and convincing arguments about humanity's vital problems: new data on the effectiveness of the atomic bomb, new frightening forecasts of population growth and the food crisis.' The individual scientist, however, had no possibility of creating change. He 'shouts a few times and is soon used up', and he is not infrequently declared a 'political idiot'. Nevertheless, Fichtelius emphasized that 'scientists in responsible positions must from time to time cry out so loudly that politicians listen for a while'. When it came to critical issues for humanity, there was no time for passivity and helpless resignation. The important thing was not more research or technological innovations – all the necessary funds were available. The politicians had 'mass media to influence the people, telephones to call their colleagues in other countries, planes to go there on, effective

---

66 *Ibid.*
67 Hans Palmstierna, 'Förskingringens sekel', *DN*, 15 January 1967; Hans Palmstierna, 'Profiten först – hälsan sen', *DN*, 17 February 1967; Hans Palmstierna, 'Förskingringen kan hejdas', *DN* 21 March 1967.
68 Letter from Hans Rabén to Hans Palmstierna, 19 May 1966, 452/1/4 (HP ARBARK).

contraceptives to implement birth control, and so on'. The route to world peace and world government lay open.[69]

Even so, Fichtelius believed that the work should begin at home. The method he proposed was a governmental information campaign about humanity's situation. He envisioned 'a never-before-seen information campaign' which could provide voter backing for an active Swedish foreign policy. With the majority of the Swedish people behind them, the politicians could turn to the UN and show what they had achieved. There they could offer to invest money in similar information campaigns in all countries. Scientists around the world were sure to support the initiative and help. 'What happens next is entirely in the hands of the politicians', he concluded.[70]

The programme Fichtelius presented aimed to bring about a new union between science and politics. The path to this fusion went by way of edification and information. He believed that knowledge and insights into the gravity of the crisis could be disseminated in a linear fashion: from scientists to politicians, from politicians to the general public, and from Sweden to the world. All that was required was determination, mass media, and financial resources. In a subsequent article, he developed his view on how politicians – 'the practical sociologists' – could bring about social change. Fichtelius supplied an example from the world of jackdaws. 'If a young, low-ranking jackdaw shows signs of unease, it is ignored by the others in the flock', he wrote, 'but if one of the leaders, the high-ranking birds, does the same, the whole flock reacts.' Psychologists had shown that humans functioned in the same way, and this was where the politicians' opportunities lay. Being at the top of the social hierarchy, they had unique opportunities to exert an influence.[71]

Throughout the summer of 1967, Fichtelius continued to apply 'biological observations to the current political debate'.[72] He particularly stressed that humans were one animal species among many. They had to overcome their pride, acknowledge their animal nature, and begin to live in accordance with the laws of biological reality.[73] However, he did not go into the question of what this would mean in practice.

---

69 Karl-Erik Fichtelius, 'Vad väntar politikerna på?', *DN*, 19 March 1967.
70 *Ibid*.
71 Karl-Erik Fichtelius, 'Biologisk sociologi', *DN*, 23 May 1967.
72 Karl-Erik Fichtelius, 'Ett underbart djur', *DN*, 8 June 1967.
73 Karl-Erik Fichtelius, 'Skapelsens krona?', *DN*, 25 June 1967.

Concurrently, Hans Palmstierna became more closely linked to *Dagens Nyheter*. At the end of May 1967, he was invited to an informal round-table discussion on current environmental issues, to be followed by dinner. The goal was to arrive at a common strategy for the continued opinion-forming process. In addition to Olof Lagercrantz and Sven Fagerberg, the participants in the meeting included the journalist Barbro Soller and the ecologist Bengt Lundholm.[74] It is noteworthy that Palmstierna wrote a letter to Lagercrantz before the meeting, inquiring about the possibility of employment at the newspaper.[75] The issue seems to have been raised before, and it shows that Palmstierna had very advanced plans to leave the research world behind. He wanted to influence people and drive social change.

The third global threat

Hans Palmstierna and Karl-Erik Fichtelius's first year in the public sphere is interesting for several reasons. First, it shows how the two scientists gradually built up a position in a key knowledge arena. They thereby became visible to more people and gained contacts which became absolutely crucial for the development of that position. Without the Malthus article and the friendship with Sven Fagerberg, *Plundring, svält, förgiftning* would not have been written. Equally important was that Palmstierna and Fichtelius built up a degree of interest around themselves over time. Not all debate books published in the autumn of 1967 were written about by the press. Even fewer featured in the evening news or were the focus of a televised cultural magazine.

A comparison with Rolf Edberg's actions at this time is enlightening. What was he doing? My research shows that he did not in fact participate in the Swedish debate. *Spillran av ett moln* was a one-off contribution. It was neither preceded nor followed by any articles or television appearances – at least, not in Sweden. From a Norwegian perspective, the situation looks different. Edberg lived in Oslo, and after ten years as ambassador he had a wide-ranging network of Norwegian contacts. Accordingly, he was interviewed on Norwegian

---

74 Letter from Ingemar Wezelius to Hans Palmstierna, 27 May 1967, 452/3/2 (HP ARBARK).
75 Letter from Hans Palmstierna to Olof Lagercrantz, 30 May 1967, 452/3/2 (HP ARBARK).

television, gave lectures, and published short excerpts from his book in various magazines. In 1966 and 1967, the Norwegian public sphere was his home ground. He usually declined invitations from Sweden.[76]

There were other differences too, of course. Edberg was a layman, not a scientist. In the autumn of 1966 he was a lone warning voice, not part of a choir. *Spillran av ett moln* was a hardback, not a cheap paperback. The most important difference, though, was that he did not personally continuously intervene in the public debate. Hans Palmstierna and the researchers behind *Människans villkor* did so. As a direct result, knowledge about a global environmental crisis began to circulate with a new intensity in Sweden in the autumn of 1967.

Additional conclusions can be drawn from the activities of Palmstierna and Fichtelius, though, not least with regard to the environment and the role of the environmental crisis in the Swedish survival debate of the 1960s. The environmental threat was not particularly prominent in that debate during 1966. For Palmstierna the focus was overpopulation, and for Fichtelius it was the nuclear-weapons menace. In the autumn of 1967, however, the centre of gravity shifted from these two established global threats to the third one. At the time, though, it was not possible to draw any sharp boundaries. Environmental and population issues in particular were intimately intertwined. They comprised an integrated complex of problems with a mutually reinforcing dynamic.

Hans Palmstierna and Karl-Erik Fichtelius undoubtedly played a decisive role in the shift which occurred in Sweden in 1966–1967. However, the social breakthrough of knowledge was not their work alone. To understand the historical process, it is important to study other actors as well: people who were not in the most glaring limelight, but who were nonetheless nearby. Consequently, the next chapter provides a different perspective on the course of events by following the journalist Barbro Soller and the historian Birgitta Odén.

---

76 Larsson Heidenblad, 'Ett ekologiskt genombrott?'

4
How the journalist and the historian came to the environmental issues, 1964–1969

In the course of the 1960s, the journalist Barbro Soller and the historian Birgitta Odén developed a deep involvement with environmental issues. As a result, their lives and careers moved to new paths. Soller took part in the establishment of a new journalistic field. Odén strove to do the same in the history discipline. However, their respective initiatives led to completely different results. Whereas Soller achieved success and renown, Odén encountered resistance that made her break off her attempts.

Barbro Soller and Birgitta Odén were prominent professional women at a time when such individuals were unusual. The worlds they worked in were dominated by men. In 1965, *Dagens Nyheter* employed 183 journalists. Only thirty of them were women.[1] In November that same year, Odén became a professor at Lund University. All her colleagues and all her predecessors were men. Odén herself did not draw attention to her pioneering role, however.[2] Nor did Soller. The two women operated through example and action rather than through debate and polemic.

As we have seen in the previous chapters, the big breakthrough of environmental issues in Sweden followed two different paths. One was apocalyptic and global; the other was more low-key and national. Soller and Odén were mainly active in the latter. While taking environmental destruction seriously, they did not talk about it as constituting a threat to human survival. Nor was their focus primarily on overpopulation or weapons of mass destruction. Tracking their actions therefore reveals somewhat different paths of development from those followed by Hans Palmstierna, Karl-Erik Fichtelius, and Rolf Edberg.

---

1 Anon., 'Ett väl sammansvetsat DN-lag', *DN*, 1 June 1965.
2 Eva Österberg, 'Birgitta Odén', in *2017 Yearbook* (Stockholm: The Royal Swedish Academy of Letters, History and Antiquities, 2017), p. 28.

My study of the two pioneering women begins with Barbro Soller's journalism. I follow her from the time she was hired as a general reporter in 1964 until she left *Dagens Nyheter* in the summer of 1969. During this period she became Sweden's first environmental journalist. But when did this happen, and how? And what role did she play in and for the breakthrough of environmental issues in Sweden? I then investigate how Birgitta Odén, in collaboration with the Swedish National Defence Research Institute (FOA), political scientists, and economists, developed the interdisciplinary project 'Miljö, naturresurser och samhälle' [Environment, natural resources and society]. I follow her from the time when the first steps were taken, in May 1967, until she abandoned the project in the spring of 1969. I also analyse how she worked at the local level to build up an environmental-history research speciality at the Department of History at Lund University. First, though, we will have a look at the late winter of 1964, when Barbro Soller shook up the Swedish biocide debate for the first time.

General reporter and environmental journalist, 1964–1966

On Saturday 22 February 1964, Barbro Soller reported that mercury had been detected in Swedish hen eggs. *Dagens Nyheter* made the news its lead story. The front page featured a photograph of the chief physician and associate professor Stig Tejning. Wearing a white lab coat and holding tweezers in his right hand, he peered gravely into the camera. The lead paragraph stated that unlike other animals, hens rarely became ill when fed mercury-treated seed. The hens' resistance posed a danger to humans: 'The mercury that the hens eat passes into the eggs we eat', said Soller. Inside the newspaper, she explained the procedure followed by Tejning and his colleague Ragnar Vestberg in detail. She was careful to let the researchers speak for themselves.[3]

The next alarm came on the following Friday. Soller reported that a new study showed that 'we Swedes also carry a dose of DDT in our body fat'. The findings came from Associate Professor Gunnar Widmark. The newspaper's front-page article explained that he was part of a large research group which had prepared a new proposal for Swedish biocidal research. The aim was 'to clarify the general situation and shed light on the ecological effects resulting from the

---

3 Barbara, 'Kvicksilver i svenska ägg', *DN*, 22 February 1964.

use of chemical pesticides'. The application had been prepared by Bengt Lundholm, who emphasized that several of the sub-projects were 'of such a nature that they must be implemented immediately'. This was especially true of the issue of mercury poisoning. 'We cannot afford to lose another year', he stressed.[4]

What Soller reported on in February 1964 was not the result of scientific studies pursued for many years. On the contrary, she wrote about preliminary studies and applications which had emerged in the wake of the intense Swedish biocide debate in 1963. Through Soller, researchers were able to present preliminary results to the public and demand more research funding. In the weeks that followed, she wrote continuously about Tejning and Vestberg's new experiments. Among other things, their studies showed that the mercury accumulated in the egg white. The levels were much lower in the yolk, and there were no traces in the shell.[5] The researchers also pointed out that hens which had eaten mercury-treated seeds did not lay their eggs in the nest boxes. This might, they said, account for Swedish birdwatchers' reports of nest failure among white-tailed sea eagles and other bird species.[6]

Barbro Soller's articles about environmental toxins attracted a good deal of attention at the time. They shifted the focus from Rachel Carson's *Silent Spring* to studies of Swedish conditions. The mercury-laden eggs were debated on editorial pages and discussed by representatives of various government authorities. Within environmental-history research, this has been taken as evidence that Soller became Sweden's first full-time environmental journalist in 1964.[7] Some nuance should be brought into that interpretation, though. At this time Soller was writing about all sorts of things, from choral singing and art exhibitions to foreign celebrities visiting Stockholm.[8] She was certainly deeply committed to environmental issues; but she was a general reporter by profession.[9]

---

4 Barbara, 'Giftkontroll i stor skala', *DN*, 28 February 1964.
5 Barbara, 'Foderexperiment med höns. Mest kvicksilver i äggvitan', *DN*, 9 March 1964.
6 Barbara, 'Värpning utanför redet kan förklara fågeldöd', *DN*, 11 March 1964.
7 Djerf Pierre, *Gröna nyheter*, p. 114.
8 Barbara, 'De tränar för konsertresa till Amerika', *DN*, 18 February 1964; Barbara, 'Tusen såg vårvernissage i DN-regi', *DN*, 9 May 1964; Barbara, 'Ungdom om Chrustjevs besök', *DN*, 25 June 1964; Barbara, 'Mera slagsmål med Floyd', *DN*, 2 July 1964.
9 The following is based on Larsson Heidenblad, 'The Emergence of Environmental Journalism'.

How journalist and historian came to the issues

This phase of her career lasted from 1964 until 1966. During this period, other topics she wrote about included Charlie Chaplin's visit to Stockholm and what lifeguards did at a west-coast seaside resort.[10] One of her specialities was lengthy articles about animal and birdlife, usually published in the Sunday supplements and illustrated by colour photographs. These texts indicate her deep interest in nature, but they were hardly a form of environmental journalism. Their focus lay on the animals' lives and behaviour. There was no discussion of any environmental crises or toxins.[11]

Over time, though, Soller's position at *Dagens Nyheter* became more and more established. In the spring of 1966, she had the opportunity to make her first major reportage trip abroad. Together with the photographer Stig A. Nilsson, she travelled around India for a month to report on the looming hunger crisis. *Dagens Nyheter* marketed the article series in advance, and all three parts featured prominently on the front page. It is worth noting that the first part was published exactly one week after Hans Palmstierna's 'Malthus och världssvälten'. But whereas Palmstierna was a scientific debater, Soller was a journalist who painted pictures. She informed readers that it was a drought year in India. The ground was ulcerous and full of cracks. River beds had been transformed into burnt valleys, and the wells had dried up. 'The soil, humans and animals are thirsting as they have not done for a hundred years.' The reason was that the monsoon rains had not come, and harvests had failed throughout the country. India depended on help from the outside world.[12]

Soller stressed that 46 million of India's 480 million people were threatened by an acute food shortage. To ensure that aid reached them, the authorities built new roads and dams. These relief efforts gave poor families a chance to make a living. They could use their wages to buy food at fixed prices. Through texts and photographs, Soller and Nilsson gave faces to the humanitarian catastrophe. One of those depicted was a family man sitting with his back to his

---

10 Barbara, 'Kärlek vid första ögonkastet när Chaplin mötte "Vasa"', *DN*, 3 November 1964; Barbara, 'Åtta livräddare vaktar Tylösand', *DN*, 1 July 1966.
11 Cf. Barbara, 'Älgar skyr moderlös kalv', *DN*, 31 March 1965; Barbara, 'Orrtuppen kuttrar året om', *DN*, 18 April 1965; Barbro Soller, 'Bild av universum medfödd hos fågel', *DN*, 2 October 1966.
12 Barbro Soller-Svensson, 'Hungersnöd hotar 46 miljoner indier. Barn bygger "svältens väg"', *DN*, 10 May 1966.

family. He had given his daily ration of bread and water to his wife and daughters. To avoid feeling hunger, he was looking the other way while they ate.[13]

The second part of the series criticized the way in which the Indian authorities were handling the crisis. Soller said that detailed maps of the famine had been drawn up. On them, it was possible to read what percentage of the population in an area had been affected by drought and crop failure. 'But if one reads the reality, one easily loses respect for the statistics', she wrote, and singled out a village which the maps claimed had been spared. The government sent purchasers there to buy hundreds of kilos of grain. But the villagers had no surplus. They therefore had to buy at a high price in the market and sell cheaply to the authorities. 'It is more than likely that the purchaser became blinded by the map statistics', Soller commented. In connection with the report, she interviewed India's minister of agriculture, whom she described as 'a feisty, energetic little man' with irrepressible optimism. His goal was to get Indian farmers to grow crops scientifically. This called for land reforms, tractors, new types of grain, chemical fertilizers, and chemical pesticides. The vision was that India would be self-sufficient by 1970. Soller was sceptical.[14]

The final part of the series dominated the Sunday edition on 15 May 1966. *Dagens Nyheter* wrote that the series of articles had aroused strong feelings in its readers. Many of them wanted to help. The paper therefore published a list of current and planned Swedish relief measures to which readers could donate money.[15] Soller interviewed and portrayed Prime Minister Indira Gandhi, but without any critical edge. Gandhi was 'a person with a difficult job' who was doing it 'with true Indian dignity'. In their conversation, the two women discussed what the West could learn from India.[16] The article about the ongoing population explosion was far more censorious. Soller stressed that India's population was growing by one million people a month, which she said was an untenable situation. All measures to improve people's living conditions 'will be eaten

---

13 Soller-Svensson, 'Hungersnöd hotar 46 miljoner indier'.
14 Barbro Soller-Svensson, 'Regeringen köper upp de fattigas "överskott"', *DN*, 12 May 1966.
15 Anon., 'Många möjligheter hjälpa "Det svältande Indien"', *DN*, 15 May 1966.
16 Barbro Soller-Svensson, 'Indira Gandhi – "en person med ett svårt job"', *DN*, 15 May 1966.

up by the rising excess population', unless 'the measures are specifically focused on family planning'. So far, though, all campaigns had been ineffective. Despite the fact that over a million sterilizations had been carried out and 16,000 family planning centres had been set up, India's population had continued to grow with undiminished vigour. By the year 2000, the country would be home to an estimated one billion people. It was a frightening prospect.[17]

Barbro Soller's role in the article series 'Det svältande Indien' [Starving India] was that of the committed reporter. She had a personal style, and she did not hesitate to deliver criticism and draw her own conclusions. In this she was part of a larger trend. The 1960s saw a shift in the journalistic ideal, from objective mirroring to critical scrutiny.[18] The shift occurred alongside the professionalization of journalism. Journalism colleges were established in Sweden in 1962, and in the course of the decade the ties between the political parties and the press weakened as well. These changes made a more independent journalistic role possible.[19] That development turned out to be of great importance for Barbro Soller and for the breakthrough of environmental issues in Sweden.

After returning from India, however, Soller returned to being a general reporter for some time. She worked in this way for the rest of 1966. Her articles might just as easily be about holidaymakers in the Swedish coastal province of Bohuslän or the Oktoberfest in Munich as about new environmental warnings.[20] However, her time as a generalist was coming to an end. That autumn she wrote an ever-increasing number of environmental and animal-focused articles, and from January 1967 onwards she wrote exclusively on those

---

17 Barbara, 'Hoppet står till spiralen', *DN*, 15 May 1966.
18 Monika Djerf Pierre and Lennart Weibull, *Spegla, granska, tolka: Aktualitetsjournalistik i svensk radio och TV under 1900-talet* (Stockholm: Prisma, 2001).
19 Stig Hadenius and Lennart Weibull, *Partipressens död?* (Stockholm: Svensk informations mediecenter, 1991); Daniel Hallin and Paolo Mancini, *Comparing Media Systems* (Cambridge: Cambridge University Press, 2004); Lennart Weibull, 'Är partipressen död eller levande? Reflexioner från ett presshistoriskt seminarium', *Nordicom-Information* 35.1–2 (2013); Elin Gardeström, *Att fostra journalister: Journalistutbildningens former i Sverige 1944–1970* (Gothenburg: Daidalos, 2011).
20 Barbara, '3600 turister i Smögenhem: Härligt vatten', *DN*, 4 July 1966; Barbro Soller, 'Bayrarna firar med fulla krus', *DN*, 29 September 1966; Barbara, 'Ny gifttyp i naturen: "Farligare än DDT"', *DN*, 23 November 1966.

topics. Barbro Soller was now Sweden's first environmental journalist. In that role, she would once again make the topic of mercury in food front-page news.

Barbro Soller and the mercury-laden pike 1967–1968

On Saturday 21 January, Soller reported that high levels of mercury had been measured in Swedish lake fish. The discovery was made by Sweden's National Institute of Public Health, which was doing a national survey of fish stocks. Soller reported that the ongoing studies had yielded worrying results. In Sweden's largest lake, Lake Vänern, researchers had found northern pike with up to 1.4 mg of mercury per kilo, far above the set limit value of 1 mg. The authorities therefore advised the public against eating Vänern fish every day. However, they stated as fact that people could 'consume fish from this and other mercury-contaminated watercourses once a week' without risk. Soller was critical. She pointed out that the limit values for other foods were set at 0.05 mg of mercury per kilo. 'So why do you recommend a value for fish that is 20 times higher?' she asked. Professor Arvid Wretlind replied that fish accumulated mercury faster than land animals, so it was necessary to allow for a higher natural baseline value. There was no reason to panic, he stressed, because the average Swede ate such small amounts of fish. The new recommendation was only intended for people with an unbalanced diet featuring lake fish.[21]

Soller was not reassured, however. She monitored studies of Swedish lake fish closely, and her reports repeatedly appeared on the newspaper's front page. In mid-February 1967, a major meeting about mercury was held at the Government Offices in Stockholm. Acting together, the Ministries of Agriculture and Social Affairs had invited six researchers, including Stig Tejning. From the meeting, Soller reported that the mercury content in Swedish lake fish was nineteen times higher than that of sea fish.[22] Suspicions fell on industry, and studies pointed the finger at the pulp mills.[23] In March, she reported on Tejning's continued research into mercury-treated seeds. The studies showed that people who worked with mercury could be irreparably harmed. Soller argued that this evidence should

---

21 Barbara, 'Giftgräns för fisken i insjöar. Bara en gädda i veckan', *DN*, 21 January 1967.
22 Barbara, 'Varning för insjöfisk', *DN*, 14 February 1967.
23 Barbara, 'Ny giftkälla kartläggs', *DN*, 16 February 1967.

be considered highly significant to 'the ongoing discussion about what mercury levels we should accept in fish for sale'.[24]

In parallel with Soller's mercury campaign, the Swedish survival debate had begun to gain momentum. Hans Palmstierna and Karl-Erik Fichtelius were regularly writing debate articles on *Dagens Nyheter*'s culture page. Rolf Edberg's *Spillran av ett moln* was about to be printed in a third edition. In April, Soller's and Edberg's paths crossed. This was the first time that Sweden's ambassador to Norway had been interviewed in the Swedish press about *Spillran av ett moln*. The prominently placed article was presented as the launch of a new series of articles entitled 'Den hotade människan' [Endangered humanity]. Soller pointed out that Edberg's book had become a bestseller in scientific circles. In the interview, she took a back seat and allowed Edberg to present his case. He maintained that the prosperity of countries was often 'measured in the number of cars, televisions, and telephones'; but in his view, the degree of environmental degradation was a far more reliable measure of 'civilization'. He personally believed that prosperity should be measured in a different way: 'Air to breathe instead of poison to breathe. Water to drink instead of poison to drink.' Edberg's pointed statements were illustrated by a half-page colour photo of a rubbish tip.[25]

The next day, a follow-up article appeared. Soller interviewed the professor of bacteriology Carl-Göran Hedén, a contributor to *Människans villkor*, who directed sharp criticism against the politicians and the political system. He said that more people with Rolf Edberg's clear-sightedness were needed. Humanity's most important issues required leaders with insight into the biological situation of the species. The problem was that today's politicians were only responsible 'to their voters and not to their voters' children and grandchildren'. Consequently, immediate and topical issues were prioritized over long-term threats. Hedén focused in particular on the population issues and the imminent global food crisis. Research, he felt, should concentrate on these issues in order to solve the general problems. Just as in her Edberg interview, Soller herself was almost invisible in the text. She asked brief questions which Hedén answered at length, and she did not express any criticism or personal

---

24 Barbara, 'Arbete med kvicksilver kan ge obotliga skador', *DN*, 4 March 1967.
25 Barbro Soller, 'Vi plundrar våra barns jord', *DN*, 9 April 1967.

views. Her role in the article series was to ensure that the warning voices were heard.[26]

After the Hedén interview, however, no more articles were published in the 'Endangered Humanity' series. This fact says something about how Soller conducted her environmental journalism at this time. She moved quickly from topic to topic and did not adhere to any set plan. In the months that followed, she wrote about such topics as oil spills, conservationists in Skåne, the restoration of silted-up lakes, and the establishment of a new educational programme in environmental engineering.[27] But the topic to which she consistently returned was the mercury content in Swedish lake fish. In the autumn of 1967, she intensified her coverage.

On Friday 13 October, Soller reported that people who ate large amounts of fish from Lake Vänern could suffer brain damage. Stig Tejning was again responsible for this alarm. He had examined 'the blood and hair of 54 extreme fish-eaters around Lake Vänern', mainly professional fishermen and their family members. On average, the group members had five times more mercury in their blood cells than a control group. Tejning stressed that the situation was extremely serious and that the fishing industry was at risk of being wiped out for reasons of national health. Commercial fishermen were facing an economic disaster. Through Soller, Tejning also expressed his hopes for additional funding so that his research work could continue.[28]

In mid-October 1967, the National Institute of Public Health completed its major geographical survey. Soller maintained that it confirmed all suspicions.[29] One month later, the National Swedish Board of Health and the National Swedish Veterinary Board met. They banned the sale of fish from about forty watercourses, but not from Lake Vänern. Soller scathingly described this decision as proof that Sweden was now the first country in the world not to follow the World Health Organization's recommended limits on

---

26 Barbara, '"Extrapoäng åt forskning om svälten"', *DN*, 10 April 1967.
27 Barbara, 'Sandhamn i uppror: Tjockolja förstör', *DN*, 17 May 1967; Barbara, 'Oro men ingen hopplöshet bland Skånes naturvårdare', *DN*, 28 May 1967; Barbara, 'Lortsjö föryngras med tusentals år: Gyttja pumpas ut', *DN*, 26 June 1967; Barbro Soller, 'Miljöingenjör "bristyrke": Räddar luft och vatten', *DN*, 11 September 1967.
28 Barbro Soller, 'Vänerfisken kan orsaka hjärnskada: Docent slår larm', *DN*, 13 October 1967.
29 Barbro Soller, 'Kvicksilvret i svensk fisk kartlagt', *DN*, 18 October 1967.

mercury content in human food.[30] In the following months, Soller reported on public debates. She also began writing editorials on the issue,[31] arguing that the set limits were arbitrary and the sales ban had to be extended. She put pressure on the authorities by means of questions and articles.

On Saturday 3 February 1968, Soller triumphantly reported that 'the National Swedish Board of Health admits it was wrong about toxic values in Swedish lake fish'. The set limit value only applied if someone ate fish just once a week (something Soller had in fact reported a year earlier). However, this information was missing from the material that the National Swedish Veterinary Board had sent out to the country's municipal health boards. Therefore, Soller was now taking matters into her own hands. With her article, she wanted to tell the Swedish people the facts. On her side she had researchers and environmental debaters such as Stig Tejning and Hans Palmstierna.[32] The close collaboration between the environmental journalist Soller and parts of the scientific research community was characteristic of the breakthrough of environmental issues in Sweden. Affording the researchers voice and publicity, she also pursued an increasingly independent and critical approach towards certain authorities and the researchers who worked for them. Her position at *Dagens Nyheter* was strengthened, and the opportunities for her to conduct investigative journalism increased.

Nya Lort-Sverige 1968–1969

Barbro Soller's big public breakthrough as an environmental journalist came in the spring of 1968 with the reportage series 'Lort-Sverige 30 år efteråt' [Filth-Sweden 30 years later]. The title was a reference to Ludvig ('Lubbe') Nordström's social reportage *Lort-Sverige* of 1938, a Swedish classic which had shown how poor people in 1930s Sweden were living in dirty, draughty homes with bedbugs, fleas, and cockroaches. The reportage shocked contemporary society and became a driver of Social Democratic reform

---

30 Barbro Soller, 'Fisk i 40-tal vatten otjänlig människoföda trots höjd giftgräns', *DN*, 15 November 1967.
31 Barbro Soller, 'Kvicksilverexpert slår larm: Förgiftad fisk kan ge allvarliga fosterskador', *DN*, 26 November 1967; Barbro Soller, 'Kvicksilverutsläppen vållar miljonförluster', *DN*, 9 December 1967.
32 Barbro Soller, 'Folkhälsan medger fel om giftvärdena för insjöfiskarna', *DN*, 3 February 1968.

policy. By 1968, however, Nordström's Filth-Sweden was history: within thirty years, Swedish homes had become clean and modern. But Soller claimed that this progress had a dark side. The dirt had not in fact disappeared at all. It had just been moved out into the natural environment.[33]

'Lort-Sverige 30 år efteråt' was the outcome of new collaboration between Soller and the photographer Stig A. Nilsson. In the spring of 1968 they travelled around Sweden, just as they had travelled around India two years earlier. Together, they revealed and investigated a variety of unsatisfactory conditions. All seven articles featured prominently on the front page of *Dagens Nyheter*, occasionally accompanied by colour photographs. Each part began by quoting a passage from Nordström's 1930s report and comparing it with the present. Soller placed particular emphasis on the environmental consequences of modern comforts such as cars, flush toilets, washing machines, and dishwashers. Among other things, she portrayed a family with small children in Värmland and asked the National Swedish Institute for Building Research to calculate the environmental impact of their life in a modern house.

In the article, Soller alternated freely between different styles. She wrote in the form of a fairy tale about how the young family had left their cramped apartment to move into their new house. It was 'a very common tale in 1968', she emphasized; 'all over the country, women and men rejoiced at the advance of technology in kitchens and bathrooms'. The machines made life easy and clean. 'But all good fairy tales contain a troll, or at least a dismal chapter', she added. So did the fairy tale about modern Swedish life in a house. She then switched to an objective tone and described how the family was indirectly polluting the water, land, and air. She gave precise figures and compared them with the situation thirty years earlier. The trend was worrying. Soller therefore appealed to her readers not to use phosphate-laden laundry detergent or install kitchen waste-disposal units. 'We are all involved in creating the environment we live in', she wrote, 'but we can improve it, and great efforts are being made to reduce everyone's contributions to the total mass of pollution.' She looked ahead with hope to a society where rubbish would become valuable compost and not air pollutants.[34]

---

33 Barbro Soller, 'Nya Lort-Sverige synas av DN: Avloppsslam på avvägar', *DN*, 26 March 1968.
34 Barbro Soller, 'Hög standard smutskälla', *DN*, 11 April 1968.

Another report in the series focused on Stockholm's urban air. The front page featured a picture of a man from the city health board. He was holding a sheet of originally white paper that had been stained black by exhaust fumes. Behind him was a mass of cars. Thirty years earlier, Soller said, there had been about 60,000 motor vehicles in Stockholm. By 1968 the figure was 360,000. Car exhaust fumes mingled with the chimney smoke from oil boilers, district heating plants, and factories. The best protection against the Stockholm air was to cough.

Soller explained that many people had become accustomed to the gradual deterioration in air quality. Others protested, however. She described how a growing group of inner-city parents had begun to lobby the politicians. The parents did not want their children to have to grow up in an environment that harmed their health. They therefore demanded an immediate ban on lead in petrol, the implementation of mandatory catalytic converters, and the revocation of plans for new multistorey car parks in the city centre. Soller predicted the formation of an 'increasingly vigilant opinion'.[35] In the spring of 1968, she herself was very much involved in the creation of that opinion.

The 'Filth-Sweden 30 years later' series of articles also contained reports about agriculture and industry. Soller contrasted the sanitary problems of lice and flies in the Swedish homes of the 1930s with the widespread use of chemical pesticides in the 1960s. She had been involved in biocide issues for a long time and had good contacts with the world of research. But in the spring of 1968, she applied a journalistic approach that was new in this context: she appealed directly to farmers. What did the people who grew and sprayed the crops really think?

The small-scale farmer Gösta Olsson in Skåne replied that, like many others, he had subcontracted his spraying to an agricultural service-supply agency – 'today's new farmer'. The year before, though, he had stopped using mercury-treated seed on his own initiative and still had a good harvest. Soller said that Olsson was not alone in taking matters into his own hands. During the 1966–1967 season, the Seed Testing Institute had found that only one-third of the Swedish hay and grain crop required treatment, and only 40 per cent of that third was actually treated. In Soller's words, the farmers had 'begun their silent opposition to seed treatment, having become

---

35 Barbro Soller, 'Mest koloxid vid trafikljus i rusningstid', *DN*, 4 April 1968.

increasingly aware that there might be consequences that would be hard to grasp'.

The next person she turned to was Henning Randau, director of one of the country's largest agricultural service-supply agencies. He was responsible for controlling weeds and pests on about 6,000 hectares of arable land in north-western Skåne. The newspaper's front page featured a picture of him wearing a full protective suit, gas mask, and gloves. In front of him stood drums of poison, and behind him the fields stretched out. He told Soller that the individual farmer could no longer keep up with developments. There were 700 to 800 chemical treatments on the market, and the equipment was expensive. As a result, agriculture had become increasingly large-scale and industrial.

Randau, however, did not become a target for criticism. On the contrary, he appeared responsible and forward-looking. He told Soller that new organic herbicides, equally effective but less toxic and easier for nature to break down, were in the process of replacing DDT. Besides, he expressed disapproval of the fact that no training was required to handle poisons. Many managers and sprayers working for agricultural service-supply agencies had only completed a short three-day course. Randau felt that that training did not amount to much. He held up the course textbook, *Kemiska bekämpningsmedel* [Chemical pesticides], the contents of which he claimed only trained chemists were able to comprehend. To young farm workers it was more confusing than clarifying. But then, there was no one with any hands-on experience on the national Poisons Board. 'When did the board last see a field?', Randau demanded.[36]

The last part of the reportage series focused on the pulp industry in Sundsvall in northern Sweden, where the country's largest forestry-industry company, Svensk Cellulosa AB (SCA), operated. Soller described how Ludvig Nordström had been impressed by the factory chimneys billowing smoke when he came to Sundsvall in the 1930s. To him, this was proof of industrial progress. To Soller, by contrast, it was a hugely polluting production process. She reported that SCA was using the sea to get rid of mercury and fibre mass without considering the long-term environmental consequences. Only in recent years had purification measures begun to be implemented. But much remained to be done.[37]

---

36 Barbro Soller, 'Bondens tysta motstånd: Obetat utsäde går lika bra', *DN*, 6 May 1968.
37 Barbro Soller, 'Ett "Lort-hav" vid Sundsvall', *DN*, 19 June 1968.

The next day, *Dagens Nyheter*'s lead editorial discussed Barbro Soller's reportage series. The writer stated that she had drawn 'a very dark picture' of the current state of environmental protection. Many of the small advances, such as the establishment of treatment plants, were only illusory improvements. In the course of her journey Soller had not found a single sewage-treatment plant that worked as it should. The editorial writer stated that the increasingly intense environmental debate of recent years had made people aware of the nature and extent of the problems. But what had those insights really led to? Only a small part of the environmental destruction was due to negligence and ignorance. The main reason was something else: it was economically profitable to pollute, and using poisons was cheap. 'We all receive a small pay-off via food prices', said the writer. The situation required sweeping changes. The new environmental awareness had to lead to 'a new willingness to take economic responsibility'. Through increased prices, fees and taxes, and reduced profits, everyone could help clean up Filth-Sweden.[38]

In the spring of 1968, Barbro Soller's reportage series reached *Dagens Nyheter*'s many readers. The following year, her readership expanded further when her reportage was published in a revised and expanded form as a paperback publication entitled *Nya-Lort Sverige* [New Filth-Sweden]. The book was abundantly illustrated by Stig A. Nilsson's black-and-white photographs, which formed visual evidence of what Soller described. The introduction stressed its documentary approach. Soller emphasized how she had 'seen, smelled, heard, and coughed' her way through the new Filth-Sweden. With her book, she wanted to spread information and create debate, and she hoped it would contribute to 'a faster resolution of difficult issues'. The environmental problems *are* difficult', she emphasized, but 'in Sweden we can afford to make an effort to fix them'. There were great opportunities to become a pioneering nation, a nation which would be able to help other countries 'where resources have been consumed and worn down even worse than here at home'.[39]

Soller's book *Nya Lort-Sverige* received favourable reviews and appeared in a second edition in 1970. Concurrently with its publication, Soller and Nilsson made one last major reportage trip together. It resulted in the article series 'Djurfabriken' [The animal factory], which was an in-depth investigation of the Swedish meat industry. The series

---

38 Anon., 'Lort och pengar', *DN*, 20 June 1968.
39 Barbro Soller, *Nya Lort-Sverige* (Stockholm: Rabén & Sjögren, 1969), p. 8.

ran from March to June, but it was not at all marketed – or attracted attention – in the same way as 'Filth-Sweden 30 years later'. For example, only the first part of the series was featured on the front page. This reportage series would mark the end of Barbro Soller's years at *Dagens Nyheter*. In the summer of 1969, she left the newspaper to work elsewhere. Her career as an environmental journalist was not over, though; in 1972 she was hired by Swedish Television to build up the television news department's environmental coverage.[40]

The big breakthrough of environmental issues which occurred in late 1960s Sweden was also a personal breakthrough for Barbro Soller herself. Through her journalism she merged science with everyday life. In the public media arena she was a unique knowledge actor who fulfilled many different functions. At first, her chief role consisted in providing scientific researchers with a voice and public attention. In her role at *Dagens Nyheter* she could make knowledge circulate and ensure that researchers' warnings were taken seriously. Over time, she personally gained an ever stronger and more independent position. It was she who saw to it that mercury-laden pike and the new Filth-Sweden ended up on the front pages.

Even so, the breakthrough of environmental issues in Swedish society did not only occur in the media limelight. Many of the people involved in the issues operated in other types of arenas. One of those individuals was the newly created professor of history Birgitta Odén. Through her archive of documents and supplementary interviews, I have been able to reconstruct how she tried to build up a new field of environmental history research in the late 1960s.[41] She did so at a time when there was no self-aware environmental history research field anywhere in the world.[42] Odén's ambition was

---

40 Djerf Pierre, *Gröna nyheter*, pp. 215–222.
41 The following is based on Larsson Heidenblad, 'Miljöhumaniora på 1960-talet?'
42 The origin of the environmental history field is usually dated to 1972, when a special issue of *Pacific Historical Review* was published. See Roderick Nash, 'American Environmental History: A New Teaching Frontier', *Pacific Historical Review* 41.3 (1972). For historiographical overviews see Richard White, 'American Environmental History: The Development of a New Historical Field', *Pacific Historical Review* 54.3 (1985); William Cronon, 'A Place for Stories: Nature, History, and Narrative', *Journal of American History* 78.4 (1992); Richard Grove, 'Environmental History', in Peter Burke (ed.), *New Perspectives on Historical Writing*, 2nd edition (Cambridge: Polity, 2001; first edition 1991); J. Donald Hughes, *What is Environmental History?* (Cambridge: Polity, 2006); Franz Bosbach, Jens Ivo Engels, and Fiona Watson,

to make the academic study of history more socially relevant and to help politicians make better decisions. But why was it that she, who until the mid-1960s had devoted herself to the state finances of sixteenth-century Sweden, wanted to tackle one of the burning issues of her day? What did she believe historians could contribute? And how did she go about building up a new field of research?

Two meetings at the Defence Research Institute, 1967

The origins of Birgitta Odén's environmental-history initiative can be found at the Swedish National Defence Research Institute (FOA). In the spring of 1967, the Institute had begun to discuss environmental issues in terms of security policy. Behind this move was Martin Fehrm, the director-general and head of FOA. At this time he was also chairman of the Swedish Natural Science Research Council, and he was one of the members of the ongoing commission of enquiry into natural resources. Fehrm had an idea that the models for systematic planning for the future that had been developed by FOA in the military field might also be put to use in the field of environment and natural resources. The realization of that idea called for social-science knowledge and expertise, however. In May 1967, he therefore invited three professors to a meeting at FOA: the political scientist Pär-Erik Back, the economist Assar Lindbeck, and the historian Birgitta Odén.

At this stage, there were no plans for a joint research project. Fehrm was primarily interested in testing established systems-theory planning techniques in the environmental field. This required social-science data plus knowledge of political decision-making processes. The three professors, though, explained that the material he wanted was not available. No relevant research had been done in political science, economics, or history. In order for systematic planning for the future to be possible, new research efforts were hence required.[43]

In the meeting, participants discussed whether they should join forces and write a debate book. They wanted such a book to make

---

Umwelt und Geschichte in Deutschland und Grossbritannien (Munich: K. G. Saur, 2006); Fabien Locher and Gregory Quenet, 'Environmental History: The Origins, Stakes and Perspectives of a New Site of Research', *Revue d'Histoire Moderne et Contemporaine* 56.4 (2009).

43 Birgitta Odén, 'Projektet Natur och samhälle', in Lars M. Andersson, Fabian Persson, Peter Ullgren, and Ulf Zander (eds), *På historiens slagfält: En festskrift tillägnad Sverker Oredsson* (Uppsala: Sisyfos, 2002), pp. 317–318.

politicians and the general public aware of the seriousness of the ongoing environmental destruction. The idea was that the book would culminate with a plea for major investment in targeted research activities. No decision was made, however, and the next meeting at FOA was not held until 27 November 1967. At this second meeting Birgitta Odén played a decisive role, because she had been given the task of compiling three memoranda on how the work should proceed.[44]

In these memos, Odén started out from the overall visions which Fehrm had presented in a memorandum that had circulated internally within FOA. There Fehrm stressed that meaningful and rational social planning required each decision to be made 'with the best possible knowledge about the consequences of the decision, but also and primarily with a clear specification of what the decision is meant to achieve'. The first step in the decision-making process was therefore to establish the relevant objective, evaluation criteria, and restrictions.[45]

The next step in the process was to identify and study the existing options for action. This would be done through plans and programmes in which costs and consequences were specified. The importance of 'hard-to-determine factors' and 'areas of uncertainty' would be highlighted, and a key role would be assigned to targeted research efforts. The aim of this research was to provide 'improved support material for future decision-making'. Birgitta Odén underlined this phrase in Fehrm's memo, and in the margin of the paragraph she noted: 'This is the only thing that history can be included in.'[46]

Odén's marginal note is explained in another document where she wrote down her views of Fehrm's memo. There she writes that 'everything in Fehrm's plan deals with forecasts. For this, history is useless'. What she believed that historians could contribute was 'knowledge about how society *has* worked – and works' in relevant respects.[47] She was ready to tackle the past and the immediate present, but forecasts did not appeal to her. Her marginal notes foreshadowed the frictions that would arise between FOA and the group of researchers.

---

44 This is evident from Birgitta Odén, 'Min föredragning', November 1967, BO 1.
45 Martin Fehrm, 'Välfärdssamhällets planerings- och beslutsfunktioner', November 1967, BO 1.
46 Fehrm, November 1967, BO 1.
47 Birgitta Odén, 'Mina synpunkter på Fehrms PM', November 1967, BO 1.

The November meeting was primarily a constitutive meeting, though, and Odén began her presentation by explaining why a working group was now being formed. She maintained that 'we are all deeply concerned about the consequences of the development of technology and prosperity' and underlined that the discussion could not be restricted to the scientific and technical aspects of the environmental problem complex. It was equally important to equip politicians with studies 'concerned with the economic-social-political side of the matter'. She pointed to three paths forward for the group: the writing of a joint publication, the drafting of a research programme, and the establishment of the group as a coordinating body. To be sure, the last-mentioned job required 'a mandate directly from the government'; but this item in her memorandum was not brought up for discussion.[48]

In connection with the proposals, Odén listed a number of topics for discussion: the group's qualifications, composition, funding possibilities, and relationships to various authorities. A more complex point of discussion on the meeting agenda was the question of whether the group should write a joint publication. Odén posed the question of whether the situation really called for one: 'Or has the situation changed after the *DN* debate, Palmstierna's book, the report of the commission of enquiry into natural resources, and the actions of the National Environment Protection Board?'[49] The question reveals the degree to which Odén and the FOA group's work was abreast of current affairs, and it shows how the Swedish environmental debate had been fundamentally reshaped within the space of six months. Knowledge and crisis insights no longer circulated in specific circles only, such as those at FOA, but were now moving with great intensity within the public sphere.

Birgitta Odén's second memorandum did function as a springboard, however. It was a discussion paper on how an interdisciplinary research programme should be initiated. The memo argued that 'the existing scientific data indicate that we are facing a critical point in the development of society'. The study of environmental problems could therefore 'not be limited to a scientific investigation'. What was required was an 'integrated research programme', which would consist of a scientific-technological part and a social-scientific part. She underlined that cooperation between these fields was

---

48 Birgitta Odén, 'PM 3', November 1967, BO 1.
49 Birgitta Odén, 'Min föredragning', November 1967, BO 1.

'[the whole] point of the group's consolidation'. The stated goal was to provide politicians with an improved basis for decision-making.[50]

The prerequisites for 'an investment with maximum return' were the central focus of Odén's second memorandum. She especially stressed the importance of the fact that the group considered '*the goal* sufficiently important to want to make a personal commitment and guide younger researchers who are working in the programme'. As we shall see, she herself put this goal into practice. She also felt that some form of authorization, an administrative management team, and adequate financial resources were required. Besides, in order for the relevant social-science research to be meaningful, general permission was needed to study 'the archival material of the civil service and the ministries'.[51] The November meeting concluded with those present deciding to proceed with the interdisciplinary initiative.

Discord between the researchers and FOA

The third meeting at FOA took place in February 1968. Beforehand, someone – it is unclear who – compiled a work plan for the group. The plan stated that 'the group's task was to create *a new model for values* in social planning by transferring systems-theory analysis used within FOA to the social sector'. However, constructing this model needed to be preceded by a research stage which would focus on 'the role of values in the decision-making process, the availability of relevant knowledge at various decision-making levels, and the relationship between the decision-making organizations' values and those of various opinion groups'. In order for the research task to be manageable, it was limited to environmental issues.[52]

Six disciplines supplied points of entry into the outlined research programme: science, technology, medicine, economics, political science, and history. Svante Odén and Hans Palmstierna, as well as technical experts working at FOA, were responsible for supplying the expertise in the first three areas. The coordination required extensive planning in close consultation with 'the envisaged recipients' of the findings. This concept primarily referred to the two directors-general: Valfrid Paulsson at the National Environment Protection Board and Martin

---

50 Birgitta Odén, 'PM 2', November 1967, BO 1, p. 1.
51 *Ibid.*
52 'Arbetsplan för gruppen på FOA', February 1968, BO 1, p. 1.

Fehrm at FOA.[53] Alongside the work plan, a slightly revised version of Fehrm's memorandum 'Välfärdssamhällets planerings- och beslutsfunktioner' [The welfare society's planning and decision-making functions] was also being circulated. It stressed the importance of maintaining 'the biological balance', and that environmental destruction must not be permitted to jeopardize future generations' room for manoeuvre.[54]

The level of ambition for the research programme was unmistakeably high. At the same time, the work plan did not provide any concrete guidance as to how the group should move from planning to research. In conjunction with the meeting at FOA, Birgitta Odén noted that various vested interests had begun to emerge, and that group members were therefore pulling in different directions: scientific research; social research; research policy and organization; and forecasting activities. To resolve the contradictions, she felt that the research programme's recipient – the National Environment Protection Board – should rule on what was desired, and then the group could be reorganized in accordance with that goal: 'it is up to each individual participant to organize that which is left over'. However, she perceived an imminent risk of 'a fragmentation of the group'.[55] The alternative was clear guidelines and a focusing of the research efforts. On the back of the paper, she wrote that the people who should make the relevant decisions were Valfrid Paulsson and Martin Fehrm, because they had 'the [necessary] contacts with the politicians'.[56] However, she crossed out this whole paragraph. Judging from other documents, it also seems that Odén was not prepared to be dictated to by FOA.

The clearest proof of this can be found in a letter dated 20 February 1968 to Erik Dahmén, professor of economics at the Stockholm School of Economics. The purpose of Odén's letter was to bring about an informal meeting on 5 March. Dahmén had been informed about the FOA group's work on environmental issues via Assar Lindbeck and Svante Odén, and they had invited him to participate. A scholar with a special interest in environmental issues, Dahmén was completing the debate book *Sätt pris på miljön* [Put a price on the environment] at this time.[57] In her letter to Dahmén,

---

53 'Arbetsplan för gruppen på FOA', pp. 1–2.
54 Martin Fehrm, 'Välfärdssamhällets planerings- och beslutsfunktioner', February 1968, BO 1.
55 Birgitta Odén, 'Mitt PM för 15/2 FOA', February 1968, BO 1, p. 1.
56 *Ibid.*, p. 2.
57 Dahmén, *Sätt pris på miljön*.

Odén explicitly writes that 'those of us who represent researchers outside FOA currently feel a great need to handle the assignment together and *without* FOA's involvement'.[58]

The informal meeting was duly held at the Stockholm School of Economics, but neither Dahmén nor Back was able to attend. However, Svante Odén was present, which shows that the dividing line went between the research group as a whole and FOA.[59] In a letter to Back, Odén expresses relief that the group has begun to take the practical planning work into its own hands. Her letter raises the issue of whether the contacts with FOA should perhaps be limited to collaboration with chief engineer Erik Moberg, who also 'does not want to toss this out quickly, but feels we should work on the matter for a couple of years'.[60] She concludes by saying that the whole situation had finally reached a state that felt reassuringly calm to her. Odén's letters indicate that the research group and FOA were working with different time perspectives. Martin Fehrm wanted quick results; the researchers wanted plenty of time.

Environmental history in Lund with a political focus

In parallel with the planning work at FOA, Birgitta Odén launched her own local initiatives. The first person she involved was Sverker Oredsson. At this time, Oredsson held a licenciate's degree and was completing his doctoral dissertation on Swedish railway policy in the nineteenth century.[61] A central theme of that policy was debates about the common good and the individual good. This theme also appeared in another area: the nineteenth-century forest issue. Odén encouraged Oredsson to explore this further, and in September 1967 he wrote a three-page memorandum entitled 'Miljövård och politik under 1800-talet' [Environmental protection and politics during the nineteenth century]. The memorandum supplies a brief account of the relevant laws, committee work, and political debates. Among other things, Oredsson pointed out that an investigation from 1868 had concluded that 'the destruction of forests contributed to the

---

58 Letter from Birgitta Odén to Erik Dahmén, 20 February 1968, BO 2.
59 Letter from Birgitta Odén to Assar Lindbeck, 22 February 1968, BO 2.
60 Letter from Birgitta Odén to Pär-Erik Back, 22 February 1968, BO 2.
61 Sverker Oredsson, *Järnvägarna och det allmänna: Svensk järnvägspolitik fram till 1890* (Lund: Rahm, 1969).

severe crop failure in the late 1860s'.[62] The sentence was underlined by Odén. Possibly she saw it as a warning – and hence useful – historical example.

The next piece of evidence to the effect that Odén had begun to involve her colleagues and students dates from February 1968. At that time, she sent a report to the National Environment Protection Board about planned and ongoing activities in Lund. The plan for doing basic research consisted of three parts: trend analysis, analysis from a history-of-ideas perspective, and opinion analysis. Responsibility for the first part rested with Sverker Oredsson, who was to deal with the political handling of the forest, water, drainage, and sewage issues during the period from 1850 to 1950. The history-of-ideas analysis of the nature-and-environment issue from 1890 to 1950 would be carried out by Ingrid Millbourn, who held a master's degree, and the opinion issues would be investigated by Associate Professor Lars-Arne Norborg. In addition, Odén informed the National Environment Protection Board that the team intended to conduct 'targeted, interdisciplinary research' into selected political decisions about nature-conservation issues and their effects on society. This work would be performed in collaboration with systems-analysis expertise.

The report also shows that the licentiate's-degree student Yvonne Bengtsson had started to explore the forest debate of the 1850s, and that four bachelor's-degree students had begun graduation-essay projects. The topics of their essays were the Conservative Party and the Shoreline Protection Act; Rachel Carson's *Silent Spring* and its reception in Swedish professional circles; Mörrumsån River and its problems; and the 1949 Forestry Act and the 1956 discussions.[63] The students were not named in the document, but it is clear from these topics – and the fact that the essay titles were explicitly mentioned in her communication with the National Environment Protection Board – that Odén considered students' essay-writing to be an integral part of the broader research project. She built up the local programme by guiding students and young researchers towards environmental-history themes.

One of the people whom Odén managed to steer in this direction was Lars J. Lundgren. He had become a teaching assistant in 1967, and shortly afterwards was encouraged to begin working towards

---

62 Sverker Oredsson, 'Miljövård och politik under 1800-talet', 27 September 1967, BO 1.
63 Birgitta Odén, 'Redogörelse', 20 February 1968, BO 1.

a doctorate. However, the choice of dissertation topic was not self-evident. The only thing he was sure of was that he did not want to pursue anything he had done before. Odén felt that Lundgren should take his time deciding, because he would be doing the work for many years. She believed it was not enough for the project to be interesting in purely scholarly scientific terms; the student had to feel for the subject and really want to explore it. Her exhortation led to a period of indecision before she requested a talk with him. When I interviewed him in 2017, Lundgren had clear memories of that discussion.

Odén began the conversation by saying that she had understood he was interested in current politics, modern music, and other contemporary topics. 'You seem to live quite a lot in the present', she said, 'and yet you are also a historian.' Lundgren agreed, whereupon Odén wondered if he might perhaps explore some current issue and its historical roots. Then 'she herself actually suggested this topic of the environment', employing the argument that 'you're out and about in nature so much, you should be interested in the environment'. The conversation aroused Lundgren's interest. He had followed the ongoing environmental debate, 'but never thought about it historically'.[64] Lundgren began to investigate the government commissions of enquiry and quickly realized that he was on the trail of his dissertation topic.

The conversation took place sometime in early 1968, and in the archival material Lundgren's name is mentioned for the first time in the project plan 'Natur och samhälle i svensk politik 1850–1967' [Nature and society in Swedish politics 1850–1967]. The plan, dating from February 1968, is in the form of an application to the state-funded Humanities Research Council. Because the application text is incomplete, and because subsequent documents refer to it as a memorandum, it was probably never submitted. The research plan shows that Odén was working along two distinct but intertwined lines at this time. On the one hand, she was preparing delimited research projects; on the other hand, she was involving young researchers so as to make them take these projects on. Her own role was that of research leader. There was no suggestion that she would do any empirical work herself.[65]

---

64 Interview with Lars J. Lundgren, 18 September 2017 (the recording is in the author's possession).
65 Birgitta Odén, 'Forskningsprojektet Natur och samhälle i svensk politik 1850–1967', February 1968, BO 1, pp. 4–7.

The aim of the planned historical research programme was 'to gain knowledge about the major lines of development within the set of problems concerned with nature and society over the past hundred years'. Particular emphasis was placed on how 'attitudes and values have evolved with regard to the obligations and rights of individuals towards society – and vice versa – in terms of natural resources and environmental problems'.[66] The two most extensively developed sub-projects were Sverker Oredsson's trend analysis of the forest issue and Ingrid Millbourn's history-of-ideas study of the political parties' ideological positions over the natural-resource issue from 1900 to 1930. Analysis of ideology formed a central approach in both projects, and Odén planned several similar studies, including one on the parties' positions from 1930 to 1960 and one that would address the contemporary (1960s) state of affairs. The third main area of the research plan, the development of opinion, also focused on the analysis of concepts and ideology. This area encompassed a study of the nature-conservation associations' opinion-forming activities and another of press opinions on the nature-conservation issue during the 1960s. In addition, there were sketches of socio-historical studies of members of nature-conservation associations and of whether increased leisure time led to a greater interest in nature conservation.[67]

In addition to the concrete project descriptions, the application text contains a three-page general introduction. Odén began it by referring to pre-industrial Sweden, where 'collisions between nature and society' had been 'relatively small and insignificant'. People had essentially lived within the framework of nature, and 'care for future generations was part of the world-view', among other things with regard to the management of forest resources. She emphasized, however, that even in the past there had been 'overexploitation of natural resources with catastrophic effects'. The ethical example above others was the classical Mediterranean world, whose soils had been depleted as a result of 'overly intensive grazing and forest destruction'. She underlined that this devastation was not due to a 'short-term scale of values' but rather to 'scientific ignorance about the long-term consequences'. In Odén's view, this example of natural-resource abuse and environmental destruction showed how important it was that 'cluelessness about the relationship between nature and society' should be dispelled.

---

66 Ibid., p. 4.
67 Ibid., pp. 4–7.

In industrialized society, this aspect was more important than ever before. It motivated the project's focus on modern history and its interest in ideologies, values, and political decision-making processes. Odén particularly stressed that 'the prosperity ideology's doctrine about the social and economic blessings of increased consumption' was insufficient because it did not take account of external effects on the environment. She argued that the unfavourable consequences of this neglect 'had only now become evident', and that it was therefore important to investigate how we had put ourselves in this situation. Had there been a lack of scientific information? Had political values been too short-sighted? What had the decision-making processes really looked like?[68]

These questions demonstrate that Odén attached decisive importance to political action. It was through political values, plans, and decisions that historical development was shaped. In line with these assumptions, historical research could benefit society by improving the basis for political decision-making. Research was particularly warranted in the environmental and natural-resource field because there the state of knowledge was so meagre. Odén believed that the reason for this was 'a phenomenon that does not appear to be interesting in the present and that was not perceived as interesting in past society either, [and has hence] not been felt to be immediately appealing as a research topic'. This state of things had changed thanks to the breakthrough of knowledge in society that had taken place in 1967. It was therefore important for historians to move into this new field, a field where their research was actually in demand. Socially beneficial historical research could not turn its back on the present: Odén felt that it should tackle and historicize current problems.[69]

Planning work intensifies

In March 1968, planning work entered a more intensive and more focused stage. The four professors involved – Odén, Back, Lindbeck, and Dahmén – had close contacts with one another and began to work seriously towards a common goal: a project application to the newly established Bank of Sweden Tercentenary Foundation, *Riksbankens Jubileumsfond* [now known under the latter name

---

68 *Ibid.*, pp. 1–3.
69 *Ibid.*, pp. 1–3.

only].[70] The practical work took place without FOA's participation, but there was no formal rupture. On the contrary, the professors applied to FOA for SEK 20,000 each (approximately SEK 170,000 in today's money) to launch their respective research activities. The applications were approved in April, and Odén used her funds to pay the hourly wages of some of the young researchers she had brought into the project. These scholars were commissioned to perform limited tasks that were of importance to the overall design of the research.

Two of these people were Hans Idén and Ingemar Norrlid, who together took a course in systems-theory analysis. In addition, they extracted excerpts from secondary sources in systems theory in order to investigate whether the method had previously been applied in practical nature-conservation policy. Another individual who became involved in the overall scheme was Kerstin Malcus. Through studies of relevant secondary sources, she would focus on the issue of the classical instances of natural destruction and familiarize herself with the environmental-history discussions that were going on in the United States. At the beginning of May, it was also clear that Lars J. Lundgren's research work had begun to gain momentum. He was funded to review secondary literature as well as printed primary sources dealing with the historical development of the water-and-sewage issue and the question of drainage. Furthermore, Arne Fryksén examined changes to the Nature Conservation Act during the 1950s, and Bo Huldt reviewed how the Riksdag handled the matter of the Baltic Sea.[71]

In early April Odén contacted Paul Lindblom, director of the Bank of Sweden Tercentenary Foundation. He replied that the Foundation was 'swamped by applications' and was therefore not prepared to make a decision about the group's plans until October. On the other hand, the Foundation would have greater financial resources to operate with at that point. However, because the application had to be considered by various consultation bodies, it

---

70 For an overview of the early history of the Bank of Sweden Tercentenary Foundation, *Riksbankens Jubileumsfond*, see Margareta Bertilsson, Bengt Stenlund, and Francis Sejersted, *Hinc robur et securitas? En forskningsstiftelses handel och vandel: Stiftelsen Riksbankens Jubileumsfond 1989–2003* (Hedemora: Gidlunds, 2004), pp. 19–30.
71 Birgitta Odén, 'Arbetsuppgifter för planering av forskningsuppgiften', March 1968, BO 1.

had to be submitted before midsummer.[72] The group took note of this, and the professors decided to write individual memoranda for the next scheduled FOA meeting on 6 May. The time frame for the programme was set at four years, starting on 1 January 1969. The grant they applied for was intended for the salaries of two to three young scholars in each of the sub-projects. In addition, the professors wanted to appoint a board consisting of the four of them, who would have funds for travel, conferences, and the purchase of books and journals.[73]

Birgitta Odén played a leading role in the work of the group. She was responsible for the internal communication, and she drafted the joint introduction. It is also clear that, by this time, she had made things start to happen in Lund. The economists in Stockholm and the political scientists in Umeå were still at an early planning stage. Prior to the May meeting at FOA, Odén circulated a three-page account of her activities. What is particularly interesting about this document is that the empirical exploratory drill holes she had initiated had already yielded results. Some of the young researchers had identified concrete historical problems which they wanted to investigate further.

One example of this is Lundgren's studies of the water-and-sewage question. He had investigated how the Riksdag had dealt with legislative issues regarding this matter during the early twentieth century. At that time, there was a clear conflict of interest between industry and farmers. The Conservatives sided with industry while the Liberals – supported by the Social Democrats – were on the side of those who were affected by pollutant emissions. However, when the proponents of a tougher policy gained a stronger position in the Riksdag in the 1920s, no changes occurred. Why did the activity fizzle out into passivity? The question was of general relevance, and Odén believed it deserved further investigation. Another overarching issue was responsibility for future generations. Here Sverker Oredsson's study of the nineteenth-century forest issue occupied a key position. Timber harvesting's long rotation periods led to conflicts of interest. How should economic expansion in the present be valued compared with future needs? The question also arose in discussions about another natural resource: ores. In contrast to the forests,

---

72 Letter from Birgitta Odén to Pär Erik Back, 10 April 1968, BO 2.
73 Letter from Birgitta Odén to Assar Lindbeck, 10 April 1968, BO 2.

these resources were finite, which set the stage for interesting comparisons.[74]

The person who had made Birgitta Odén interested in the ore issue was the Social Democratic politician and former finance minister Ernst Wigforss. She had interviewed him in April 1968, and he had then asserted that the Conservatives and Social Democrats had reached agreement on the particular issue of ore. Both groups believed that 'ores should be exploited at this point in time – without considering the future – because people know what they can get for the ore *now*, whereas in the future the value could fall.'[75] Wigforss said that the Liberals had opposed this position and defended the right of future generations to use these resources. The conversation with Wigforss shows that Odén did not hesitate to contact politicians. The same resolve is evident in her presentation of the planned Baltic Sea project. Bo Huldt's initial studies had shown that three specific politicians of different political colours dominated the Riksdag debates. In light of this, Odén had established contacts with them for more detailed investigations. In her report to FOA, she also mentioned five student essays which she hoped would be ready by the autumn. Their stated purpose was to explore whether some key problems in Swedish environmental policy might be suitable as case studies within the broader research programme.[76]

The economists' memorandum was not completed until after the meeting at FOA. In a letter to Erik Dahmén dated 14 May, Odén thanks him for that memo and informs him that she will 'use the weekend to complete the introduction and then send it to you all for consideration'.[77] The text Odén sent out was four pages long and she received a quick response from Back, who felt it was excellent. He had 'no amendments, not even about formal aspects'.[78] Lindbeck thought the introduction was 'sound' but reacted strongly against the passage where Odén described the origins of the group. The first draft stated that Martin Fehrm 'was one of the first to clearly realize the social-science side to this issue'. Lindbeck had crossed this out and written in the margin: 'Ugh! Servile!'[79] Erik Dahmén's

---

74 Birgitta Odén, 'Redogörelse vid sammanträde på FOA 6/5', 2 May 1968, pp. 1–2, BO 1.
75 Ibid., p. 2.
76 Ibid., p. 3.
77 Letter from Birgitta Odén to Erik Dahmén, 14 May 1968, BO 2.
78 Message from Pär-Erik Back to Birgitta Odén, 22 May 1968, BO 2.
79 Assar Lindbeck, 'Marginalkommentarer', May 1968, BO 1.

comments were mainly about formal matters, but the changes he proposed were not insignificant. In the draft, Odén had written that the government enquiry into natural resources 'had exposed frightening perspectives' and that the subsequent environmental debate 'had assumed avalanche-like proportions'. At Dahmén's suggestion, this was changed to 'had exposed grave concerns' and 'had become very lively'.[80] The changes show that Odén was strongly emotionally involved in the environmental issues, but that she was at the same time sufficiently free from prestige to change her wording. She followed Dahmén's and Lindbeck's suggested amendments at all levels.

The application, the rejection, and the research group called Natur och samhälle [Nature and Society]

On 17 June 1968, the complete application was submitted to the Bank of Sweden Tercentenary Foundation. In their preface the professors stated that, in their view, their planned research was well in line with the Foundation's stated aim of 'increasing knowledge about the effects that technological, economic, and social changes cause in society and in individuals'.[81] The application consisted of four parts. The first contained the joint introduction plus a description of the board's tasks and budget items. Three rather different presentations of the sub-projects followed. Whereas Odén had a detailed and reasoned text of nine pages, Back had submitted a more sketchy one of three. Thematically, however, history and political science were close to each other. What was to be examined was the political decision-making process and the role that values played in it.[82]

The economic project was of a different character. For Dahmén and Lindbeck, empirical studies were not enough – they aimed at theory development. The main problems to be investigated were the exploitation of natural resources, external effects, and economic planning. Their aim was to formulate a 'theory of investment in conditions of uncertainty' which assigned particular importance to 'certain special properties on the part of the relevant natural resources,

---

80 Erik Dahmén, 'Marginalkommentarer', May 1968, BO 1.
81 Pär-Erik Back, Erik Dahmén, Assar Lindbeck, and Birgitta Odén, 'Ansökan till stiftelsen Riksbankens Jubileumsfond om stöd till forskningsprogrammet Miljö, naturresurser och samhälle', 17 June 1968, p. 2.
82 *Ibid.*, pp. 13–27.

e.g. irreversibilities'. The research would be carried out in close collaboration with scientists who could help the project leaders identify suitable fields of study. The plan was five pages long, but dense in content and relatively concrete. For example, it specified which researchers were to be employed and what their qualifications were.[83]

However, the historical sub-project 'Natur och samhälle i svensk politik 1850–1965' [Nature and Society in Swedish politics 1850–1965] was the most developed one. Odén had built on her draft from February and now presented a coherent research plan. The introduction's presentation of the historical background and the purpose statements were basically unchanged, but the research design had been refined and streamlined. In June 1968, it was clear that the so-called 'trend studies' formed the core of the historical research programme. Now six in number, they may be roughly divided into two thematic blocks. The first one revolved around the political management of natural resources (forests, water, and ores); the second dealt with opinion formation and government-led nature-conservation efforts. All the trend studies aimed to 'define the lines of reasoning – i.e. with what goal and proceeding from what value – that have been employed at various times in attempts to solve problem complexes involving nature and the environment'. This objective would be achieved by means of 'the usual historical method, expanded with the quantitative method and so-called content analysis'. Because many researchers would be investigating the same material, some centralized excerpting would be undertaken, particularly of newspaper materials.[84]

The second block of trend studies would examine the emergence, social composition, and ideological development of the nature-conservation movement. Particular emphasis would be placed on the transition from 'aesthetically motivated nature conservation to socially and economically motivated nature conservation'. An adjoining study would explore the emergence of the governmental nature-conservation administration. The last trend study, of press opinions about nature-conservation issues in the 1960s, was an 'almost self-evident part of the study', but it would not be performed until 'we have attained a better distance to the topic'. In anticipation of that stage, however, a number of exploratory student essays would be written.[85]

---

83 *Ibid.*, pp. 29–33.
84 *Ibid.*, p. 18.
85 *Ibid.*, pp. 16, 17, 19.

The trend studies that were justified in the greatest detail were the forest issue and the water-and-sewage issue. In both cases, there were obvious grounds for conflict and ideological disagreements between various groups. For that reason, the intention was for the surveys to be able to contribute general insights. The forest issue was described as an 'extremely important test instrument for ideological disagreements over natural-resource issues', and the water-and-sewage issue was 'worthy of great attention, because it exhibits several characteristic features'.[86] The wording shows that Odén was not looking at historically specific characteristics but pursued general conclusions. The goal was to achieve knowledge that would be applicable in the present. Consequently, it was vital to study situations and processes that resembled current ones.

In addition to the trend studies, there was a section on historical case studies. These would act as a common resource for the research programme. If the social scientists, or FOA, needed historical expertise, special interdisciplinary studies could be conducted. At the moment of writing, plans were underway for a study about the Baltic Sea and one about drainage. The case studies would adhere to a special eleven-point template that had been developed in consultation with FOA. In this way the research could be directly useful for prognostic activities.[87]

In the project budget, Odén also discussed the forms of the collective research work. She stated that this work would mostly be done in the form of licentiate studies – i.e. postgraduate work for a degree level above the master's but below the doctorate – funded by scholarships. It was, however, 'necessary to attach more permanent staff to the project in order for it to be implemented in a more energetic and purposeful manner'. Odén wanted to do this by hiring a research assistant with a licentiate's degree who could be responsible for leading and planning the group's activities. This person would also write a doctoral dissertation on a topic related to the project. Furthermore, she pointed out that 'this kind of historical research requires relatively comprehensive and centralized collection work in various source compilations'. This work would be carried out by students on an hourly wage, but 'control of the excerpting and the excerpters should be the responsibility of a research assistant'. This individual would be a young researcher who would do a

---

86 *Ibid.*, pp. 14–15.
87 *Ibid.*, pp. 18–20.

licentiate's degree within the project's framework. Odén said that suitable individuals were available at the Department of History, but she did not mention names. As access to licentiate scholarships was uncertain, she also wanted some leeway to redistribute the grant according to need.[88]

The historical research programme that Odén had carved out was very much a collective initiative in which young researchers would form a core. Within the space of a year, she had moved from discussing the need for social-science and historical research into environmental issues at FOA to initiating a local research environment in Lund. The close contacts with social scientists, natural scientists, politicians, and the authorities meant that the research had the potential to make its mark far outside the field of history. And then, interest in environmental issues had increased sharply in Sweden during the previous year. However, in order for her plans to be realized, funding was necessary. Her hopes were focused on the Bank of Sweden Tercentenary Foundation.

At the October 1968 meeting, however, the application was tabled. From Odén's archival papers, it appears that the Foundation was 'dissatisfied with the design of the economic and political-science sub-projects'.[89] However, in the early 2000s Odén personally researched the archives to find out why the application had not been approved. She discovered that the external experts had made favourable recommendations, but that the board had not followed them. She could only speculate about the reasons for this. Even so, she had a clear recollection of a brusque statement made in the autumn of 1968 by the Foundation's secretary at that time to the effect that she should not imagine 'that historians should receive money equivalent to a full-time university lecturer's position in order to study such a subject as the environment'. She describes the message as 'a slap in the face' and points out that Martin Fehrm at FOA had felt the same way: 'To some extent, of course, what was rejected was his project, his ideas.'[90]

However, the application was tabled, not rejected; and in January 1969, Back, Dahmén, and Lindbeck submitted expanded project plans. It is apparent from these plans that the social scientists had begun doing serious work on their research ideas during the autumn

---

88 *Ibid.*, pp. 20–22.
89 Letter from Pär-Erik Back to Birgitta Odén, 9 November 1968, BO 2.
90 Odén, 'Projektet Natur och samhälle', pp. 325–326.

of 1968. Pilot studies had been implemented, and hopes for a large grant had not been dashed. Odén, however, was 'despondent after the first outcome', and she made no changes to the historical sub-project.[91]

At the Foundation's next meeting on 14 February 1969, the application was tabled once more. Contemporaneous letters reveal that Odén had now completely abandoned her hopes for external funding. 'This means the end of our group', she writes; 'neither Back nor Dahmén can hold their groups together after a year of promises – social scientists are in demand and cannot be expected to live on air.' The young researchers in Lund were 'steadfast', but Odén doubted her own role and future. 'How long will I be able to carry the load without any help in the form of a secretary or assistant?' She announced that she would scale down her activities to a low level, thereby avoiding 'any obligation to deliver anything'.[92]

An unexpected decision was announced in mid-April 1969. The Bank of Sweden Tercentenary Foundation had decided to provide funding for the economic part of the project. That announcement simultaneously meant the definite end of the group's joint venture. FOA submitted a letter of protest, but Back and Odén were not prepared to fight on.[93] Still, Odén was anxious that the environmental-history initiative in Lund should not go to waste. She felt that some form of continuation was necessary, especially in view of the young researchers she had involved. In consultation with Sverker Oredsson, she concluded that the research group 'Natur och samhälle' should be formed. However, she herself took a step back and handed over the leadership to Oredsson. During the early 1970s, the group met regularly about once a month; but this was not a coordinated research programme. The dissertation topics were rather diverse, and the group members' working conditions varied widely. Only Lars J. Lundgren and Rune Ivarsson did full-time research. In 1974, Sverker Oredsson left the research field in order to take on assignments in municipal politics.[94]

The circumstances of Birgitta Odén's relinquishing of the nature-and-society project are not entirely easy to explain. From the archival

---

91 *Ibid.*, p. 326.
92 Letter from Birgitta Odén to Jan Zeilon, 18 February 1969, BO 2.
93 Letter from Carl Gustav Jennergren to Birgitta Odén, 13 May 1969, BO 2.
94 Birgitta Odén, 'Rapport', 8 January 1970, BO 2. Interview with Lars J. Lundgren, 18 September 2017. Interview with Sverker Oredsson, 14 December 2017.

material, as well as from my interviews, it appears that she continued to be very interested in environmental issues and environmental history. It was not until the 1980s that she seriously tackled the issue again, though, and then it was from a didactic perspective. Later, she also participated in the environmental-history conferences that began to be held in Sweden in the 1990s. Lars J. Lundgren especially remembers a speech she gave at a conference dinner in the early 2000s, a speech in the course of which she looked back at her early research ideas and described her decision to switch topics in drastic terms: 'I dropped out of the group. It was better to hand it over to someone else. Because I had failed so badly and been declared an idiot.' The strong words made an impression on Lundgren, who only then realized how hard Odén had taken the rejection. Some time after the conference, he therefore asked her to expand on her thoughts, and she told him: 'I was young, I was new, and I wanted to invest my energy in a new field.' The negative decision had been too much: 'I couldn't handle it.'[95]

Odén's powerful feelings also emerge in a backward look at the project which she wrote in 2002. The text proudly states that Per Eliasson had just successfully defended 'a forest-history dissertation within the field of history, employing patent interdisciplinary approaches'. The dissertation signified that 'the ignominy from 1968/69 was washed away' and that 'environmental history in its Lundian, politicized form' could confidently proceed. Odén added: 'Ideas can be impeded by a lack of resources. But they do not have to die. They can return with new bearers and be stimulated by new impulses from the many disciplines that have the environment on their agenda.'[96] The venture had not been in vain.

Knowledge actors and networks

The histories of Barbro Soller and Birgitta Odén give us a deeper understanding of how the breakthrough of environmental issues in Sweden happened. Their professional commitment underlines that the social knowledge breakthrough involved and activated many different types of knowledge actors. In Soller's case, her interest in environmental and nature issues was accentuated. Over the course of the 1960s, she increasingly became a driver of

---

95 Interview with Lars J. Lundgren, 18 September 2017.
96 Odén, 'Projektet Natur och samhälle', p. 332.

the development. At times it was she who set the agenda of the Swedish environmental debate. In Odén's case, the breakthrough of environmental issues led to a radical reorientation of her own research – from sixteenth-century nation-state finances to modern industrial society's interaction with nature. By adopting this approach, she attempted to bring the field of history closer to the present day and to make it practically useful in social planning. However, these great ambitions did not materialize. Her own career took other paths.

Even so, Odén's environmental-historical initiative was not without results. Several of the students and young researchers she recruited would continue down the route she had staked out. This was particularly true of Lars J. Lundgren, who defended his dissertation 'Vattenförorening i Sverige 1890–1921' [Water pollution in Sweden 1890–1921] in 1974. It was followed by a long career at the National Environment Protection Board, a career which was combined with writing about environmental history. That would not have happened without Odén. This example underlines the fact that a social knowledge breakthrough is not an abstract phenomenon. A highly concrete historical process, it entails people trying new things, which then leads other people to do new things. Chains of events and lives intersect. No one is an island.

In this context, the extensive personal networks to which knowledge actors belong play a decisive role. Tracing the paths of Soller and Odén has enabled some of these networks to become visible. The two women's good relations with scientific researchers were particularly important. Soller was highly trusted as a conveyor of new discoveries. For the mercury researcher Stig Tejning, she almost served as a mouthpiece. With Soller's help, his pilot studies and experiments became front-page news, which was important in securing large research grants. Birgitta Odén's most important relationship was with her brother, Svante. He was a direct link to the research front, and he was also present at the meetings at FOA. The social knowledge breakthrough occurred in and through this type of network. In 1960s Sweden, science, politics, and the media were closely intertwined. Historical actors who understood how to take advantage of that could make a lot happen.

The clearest example of this is Hans Palmstierna. In this chapter, as in the previous ones, we have seen how his name appeared in all kinds of contexts. Palmstierna's wide-ranging social contacts are seen in his extensive private correspondence as well. After the great breakthrough in the autumn of 1967, people from all over the country, with various backgrounds and jobs, wrote to him. I have singled out a few individual examples, such as the author Sven Fagerberg and the layman Sören Gunnarsson. The next chapter applies a comprehensive approach to Hans Palmstierna's preserved correspondence as a whole, with a view to presenting a clear picture of the Swedish society in which the breakthrough of environmental issues occurred. What did people who became aware of the environmental issues actually do? Along what lines did they think about their new insights? And on what sorts of issues did they consult the foremost environmental debater in the country?

5
The environment and the Swedish public, 1967–1968

On 4 December 1967, a group calling itself 'Studentgruppen för främjandet av Naturriktig Kultur' [the Student Group for the Promotion of Nature-compatible Culture] contacted Hans Palmstierna. This newly founded student association sought to prevent the 'plundering and poisoning of nature' by arranging lectures, writing circular letters, and commenting in the daily press. The group's chair, Ulla-Britt Bergman-Holmstrand, wondered if Associate Professor Palmstierna could come to Uppsala some evening in February to give a lecture. Perhaps it might even be possible to arrange a podium debate?[1]

That same day, another Uppsala student named Berth Lundberg wrote with a similar request. He represented the city's Christian student associations, which were planning to conduct a panel debate on environmental destruction. The focus would be on biocides, air and water pollution, and urbanization and stress. In addition to inviting environmental-protection experts, the plan was to invite a theologian, who could 'incorporate the Christian idea of stewardship into the context'. Lundberg stressed that the event was not intended as 'information from experts' but instead aimed at 'normative assessments and opinion formation'.[2]

A week or so earlier, a third student association in Uppsala, the Liberal Club, had contacted Palmstierna, asking if he could come and talk about 'the destruction of nature'. The size of the audience was 'hard to estimate'; but because 'with you as the initial speaker' it would be an 'extremely attractive event', they could count on about

---

1 Letter from Ulla-Britt Bergman-Holmstrand to Hans Palmstierna, 4 December 1967, 452/3/2 (HP ARBARK).
2 Letter from Berth Lundberg to Hans Palmstierna, 4 December 1967, 452/3/2 (HP ARBARK).

fifty people.³ The Christian students thought they could attract about a hundred.⁴ Palmstierna replied that he would be happy to come to Uppsala and talk, but suggested that the three student associations hold a joint event. At the same time, he also accepted invitations from Studentaftonutskottet [a high-profile student society arranging evening lectures] in Lund, the Department of Education at Stockholm University, the Swedish Society of Psychologists, and a local Rotary club.⁵ This attention from the general public filled him with confidence. On 28 December 1967, he wrote to an American organization called Scientist and Citizen to describe everything that had happened and was about to happen in Sweden, saying, '[t]his last year has been the year when the tide has been turning, I think'.⁶

The preserved correspondence from which the above examples were taken is a unique phenomenon. Compared to the collections of other early Swedish environmental debaters, such as Rolf Edberg, those of Hans Palmstierna are of a completely different scope and character. They make it possible to visualize and analyse significant parts of the grassroots activity that followed the major breakthrough of environmental issues in Sweden. From this material, a motley group of actors, organizations, ideas, and initiatives emerges. The letters hence demonstrate that knowledge about the environmental crisis was circulating in Swedish society. Environmental issues were not only being discussed at the Government Offices and the Karolinska Institute. Interest among young students was strikingly high. Both as individuals and as members of associations, actors from this group heeded the warnings of scientists at an early stage. The students were also keen to take the step from knowledge to action. True, their financial resources were modest; but they had plenty of time and a high level of commitment.

---

3 Letter from Jan Carlsson to Hans Palmstierna, undated 1967, 452/3/2 (HP ARBARK).
4 Letter from Berth Lundberg to Hans Palmstierna, 4 December 1967, 452/3/2 (HP ARBARK).
5 Letter from Stig Lindholm to Hans Palmstierna, 5 December 1967, 452/3/2 (HP ARBARK); Letter from Kerstin Allroth to Hans Palmstierna, 11 December 1967, 452/3/2 (HP ARBARK); Letter from Klas Güettler to Hans Palmstierna, 12 December 1967, 452/3/2 (HP ARBARK); Letter from Vällingby Rotary Club to Hans Palmstierna, 28 December 1967, 452/3/2 (HP ARBARK).
6 Letter from Hans Palmstierna to Virginia Brodine, 28 December 1967, 452/3/2 (HP ARBARK).

Other actors had greater resources at their disposal. Hans Palmstierna himself was an active Social Democrat, which meant that many organizations with ties to social democracy appear in his correspondence. The most important of these was the cooperative insurance company Folksam, at that time Sweden's largest insurance company. Folksam was just as self-evident a part of the labour movement as trade unions and educational associations. In December 1967, Folksam's youth council decided to launch a nationwide campaign called 'Front mot miljöförstöringen' [A front against environmental destruction]. Palmstierna was commissioned to record films, prepare study materials, and run the campaign. This initiative was the first large-scale attempt in Sweden to turn the growing commitment to the environment into a popular movement.[7]

What also emerges from Palmstierna's correspondence is that the social breakthrough of knowledge functioned like a chain reaction. One person's action led other people to do things, which in turn made more people act. Chains of events like these are, of course, impossible to map in their entirety. In my view, however, they are absolutely crucial when it comes to understanding what a social breakthrough of knowledge is and how it occurs.

Chronologically, this chapter covers December 1967 to mid-October 1968. During this period Hans Palmstierna was constantly on the move. In March, he left the Karolinska Institute for a newly established position at the National Environment Protection Board. That summer, he became head of the Board's research secretariat. In October – a few weeks after the Social Democrats' historic victory of 50.1 per cent of the votes in the election for the Second Chamber [*andra kammaren*, the lower house of the Riksdag, which was bicameral until 1970] – he was hired by the Environmental Advisory Council to work directly under the government. Concurrently, he was a member of the group that was preparing the Social Democrats' first environmental programme; he travelled all around the country giving lectures; and he participated diligently in the press, radio, and television. However, my aim here is not to map out his activities in detail. Instead, I am using the correspondence – and to some extent press clippings – to gain a picture of everything that was happening around him within a Swedish society which realized that the environment was under threat.

---

7 Anon., 'Stor ungdomsgiv mot miljöförstörelse', *DN*, 14 December 1967; Anon., 'Front mot miljöförstöringen i ny ungdomsgiv', *Folksam: Organ för kooperativa fackliga försäkringsrörelsen (Folksam)*, 1 (1968), pp. 4–5.

The diversity of the commitment to the environment

On 2 January 1968, Sören Gunnarsson wrote to Hans Palmstierna to remind him that in the previous autumn they had discussed founding an activist group to which 'battle-hungry names' could be attracted. 'What should we do now?' he wondered. 'Can I help?'[8] In reply, Palmstierna described the upcoming Folksam campaign which he was to lead. He explained that 'Folksam has direct, unfettered ties right out into the youth movements, the cooperative movement, ABF [the Workers' Educational Association], and much more'. This was a 'huge complex' whose extent he was only now beginning to grasp. He encouraged Gunnarsson to travel to Stockholm when he had an opportunity, preferably on 30 January when the campaign would be launched.[9] This exchange reveals how fast ambitions were growing. The small-scale activist group envisioned by Palmstierna at the end of November 1967 had become outdated by January 1968.[10] Now he wanted to create something much bigger: a popular movement. Nor was he alone in that aim.

In the province of Värmland, a society called 'Vänner av Vänern' [Friends of Lake Vänern] had been formed in the autumn of 1967. Through its campaign entitled 'Rädda Vänern' [Save Lake Vänern], the society had gained significant media attention and rapidly increased its membership. At the beginning of January, it had about 1,000 members. Its chair, one Mr E. Eriksson, emphasized that the group 'had not only spread propaganda and shaped public opinion' but had also lodged three complaints with the Västerbygden Water Court. The society's activities were geared to stopping the contamination of Lake Vänern, repairing the damage already done, and bringing about stricter legislation. Eriksson told Palmstierna that the press described the society as 'a nascent popular movement which is vigorously asserting its ideas'. However, the society was burdened by financial problems. To persuade Palmstierna to become actively involved, Eriksson enclosed a number of press clippings demonstrating its media impact.[11]

---

8 Letter from Sören Gunnarsson to Hans Palmstierna, 2 January 1968, 452/3/3 (HP ARBARK).
9 Letter from Hans Palmstierna to Sören Gunnarsson, 14 January 1968, 452/3/3 (HP ARBARK).
10 Ibid.
11 Letter from E. Eriksson to Hans Palmstierna, January 1968, 452/3/3 (HP ARBARK).

The 'Friends of Vänern' initiative was the largest of its kind at this time; but many other associations were also active, albeit on a smaller scale. In the autumn of 1967, the limnological society Societas Aquatica Lundensis instituted two prizes in Lund: *vattenklövern* and *lortmedaljen* [the water clover and the filth medal]. The former was awarded to Hans Palmstierna for his successful information activities; the latter was given to the City of Gothenburg, which pleased Palmstierna very much. He characterized the filth medal as 'wonderfully infuriating'.[12] Another campaign that was just being launched focused on limiting the birth rate. The initiative came from the World Federalist Movement, part of the peace movement. In early 1968 it launched a Sweden-wide fundraising campaign for the International Planned Parenthood Federation. One of the people involved was Karl-Erik Fichtelius, editor of *Människans villkor*. He felt that, over the past year in Sweden, the ground had been 'quite well prepared for some practical initiatives'.[13] Palmstierna contributed by signing a petition, but he declined a request to come to Lund to speak to the peace-political association PAX about environmental destruction.[14]

As organizations, the 'Friends of Vänern' and the World Federalist Movement were highly dissimilar. Whereas the former was interested in the local environment and worked to influence municipal politicians and business leaders, the latter was involved in global politics. Both saw an ally in Hans Palmstierna, but they responded to different aspects of his message. Palmstierna himself moved freely between the various levels. While taking a lively interest in local and national issues, he also sought to establish international contacts, for instance with the American researcher and environmental activist Barry Commoner.[15] He was also keen to get *Plundring, svält, förgiftning* translated into other languages.[16]

---

12 Letter from Hans Palmstierna to Societas Acquatica Lundensis, 10 January 1968, 452/3/2 (HP ARBARK).
13 Letter from Karl-Erik Fichtelius to Hans Palmstierna, 8 January 1968, 452/3/3 (HP ARBARK).
14 Letter from Björn Hammarberg to Hans Palmstierna, 2 January 1968, 452/3/3 (HP ARBARK); Letter from Hans Regnéll to Hans Palmstierna, 18 February 1968, 452/3/3 (HP ARBARK); Letter from Hans Palmstierna to Hans Regnéll, 25 February 1968, 452/3/3 (HP ARBARK).
15 Letter from Barry Commoner to Hans Palmstierna, 15 January 1968, 452/3/3 (HP ARBARK); Letter from Hans Palmstierna to Barry Commoner, 29 January 1968, 452/3/3 (HP ARBARK).
16 Letter from Jacques de Reus to Hans Palmstierna, 17 January 1968, 452/3/3 (HP ARBARK); Letter from Petter Åkerman to Hans Palmstierna,

Some politicians approached Palmstierna directly in order to describe how they had begun to work with environmental issues. The Social Democrat Samuel Strandberg sent over a statement he had made to Stockholm City Council, and Axel Jansson, a member of the Left Party [in those days 'the Left Party – the Communists'], enclosed a motion his party had raised in the Riksdag.[17] Palmstierna also accepted a request from the Left Party's youth association to come to Café Marx in Stockholm and give a lecture.[18] But the political left was not the only actor that courted him. In late January, he had lunch at Handelsbanken's head office with its chairman of the board, Tore Browaldh (Handelsbanken was and remains one of the largest commercial banks in Sweden). The meeting was followed by amiable thank-you letters and a continued exchange of ideas. Browaldh was delighted to have met Palmstierna in person, the man whose book had functioned as 'something of an alarm clock' for his 'entire family'.[19]

Another Handelsbanken correspondent was lawyer Gustaf Delin. He chaired Sigtuna Town Council and represented a small group of individuals with a commitment to municipal politics. They appealed to Palmstierna for moral support and with a view to investigating whether Sigtuna could be made an 'experimental site for good environmental protection'. The group found it 'highly unsatisfactory to have to deal with environmental problems only when they become urgent'. Consequently, they wanted a system to be developed that would enable long-term planning of the municipality's activities. Delin told Palmstierna the group had recently submitted a motion to upgrade the local waste-water treatment plant. Might Associate Professor Palmstierna possibly have the time and interest to meet over lunch?[20]

---

24 January 1968, 452/1/4 (HP ARBARK); Letter from Hans Palmstierna to Jacques de Reus, 23 January 1968, 452/3/3 (HP ARBARK).

17 Letter from Samuel Strandberg to Hans Palmstierna, 9 January 1968, 452/3/3 (HP ARBARK); Letter from Axel Hansson to Hans Palmstierna, 22 January 1968, 452/3/3 (HP ARBARK).

18 Letter from the Left Party's youth association to Hans Palmstierna, 24 January 1968, 452/3/3 (HP ARBARK).

19 Letter from Hans Palmstierna to Tore Browaldh, 29 January 1968, 452/3/3 (HP ARBARK); Letter from Tore Browaldh to Hans Palmstierna, 2 February 1968, 452/3/3 (HP ARBARK).

20 Letter from Gustaf Delin to Hans Palmstierna, 18 February 1968, 452/3/3 (HP ARBARK).

A more informal group which made its voice heard in early 1968 consisted of Bo and Birgitta Wrenfelt and some friends of theirs. They had met in the couple's home for an evening of discussion about problems and environmental destruction in developing countries. The group members felt that the established political parties' handling of environmental issues was 'completely unsatisfactory'. They particularly criticized government minister Krister Wickman's assertion in the television programme *Monitor* about the electorate's lack of interest and commitment. 'How do the politicians know this?' they asked. 'As far as we know, no political party has presented its intentions regarding environmental issues before any election.' The group stressed that 'the electorate's *behaviour*' could certainly be observed – for example, from the fact that both the price of and demand for lake fish in the Stockholm area had fallen by 50 per cent in a short period of time. This could not be due to anything other than information and growing commitment. The group added that there was 'a large group of people who are seriously concerned about their own and their children's future'. They therefore wrote to Radio Sweden to request that it organize a televised debate without a set time limit, a debate which was to take place as soon as possible and include all political parties as well as leading environmental experts. Signed by seventeen people, the letter was sent to Hans Palmstierna, Radio Sweden, the five political party offices, Olof Lagercrantz at *Dagens Nyheter*, and three of the researchers behind *Människans villkor*.[21]

The many initiatives taken at the beginning of 1968 show that environmental issues had a broad public appeal. Within three months, *Plundring, svält, förgiftning* had sold 16,000 copies and been published in a third edition.[22] The letters also reveal that Hans Palmstierna was in demand. People wanted to meet, talk to, and listen to him as well as obtain his signature and support. Several editors contacted him too, asking him to write for their periodicals. While generally accommodating to editors, he largely recycled his own texts. Brief and lightly reworked excerpts from *Plundring, svält, förgiftning* appeared in many different contexts in 1968. He did continue to publish newly written texts, however, mainly in *Dagens Nyheter*; but what particularly characterized this phase was that he began to employ new media

---

21 Written communication from Bo and Birgitta Wrenfeldt, February 1968, 452/3/3 (HP ARBARK).
22 Advertisement for *Plundring, svält, förgiftning*, *DN*, 31 January 1968.

and channels. Recordings for Skol-TV [school television] and the collaboration with Folksam were of particular importance.[23]

'A front against environmental destruction'

On 30 January 1968, Folksam organized a start-up conference in Stockholm to launch the above-mentioned 'front against environmental destruction' campaign. Valfrid Paulsson and Hans Palmstierna each gave an introductory speech. The main focus of the conference was to present the design of the campaign. Folksam announced that it was chiefly aimed at young people in Sweden and was intended to function as a three-stage rocket. First, youth associations and school classes would be informed by watching a still film and working with study materials. In order 'to achieve the greatest possible activity and the best possible results', the material contained competitive elements. The young participants were encouraged to investigate their local environmental situation for themselves. They would then present the results in a creative format, such as a wall poster, an essay, or a photo montage. These could be submitted to Folksam to be judged by a jury which included Paulsson and Palmstierna. The best entries would be rewarded, the prize money totalling SEK 20,000.[24]

The goal of the Folksam campaign was to create an informed and impatient body of opinion that could drive developments forward. The offensive was to culminate in the spring of 1969, with public hearings held throughout Sweden. At these, educated and committed young people were to challenge municipal politicians, Riksdag members, business leaders, and nature conservationists. Folksam described the events as 'a kind of committee hearing'.[25] Barbro Soller compared them to the fierce interrogations that US presidents sometimes had to face.[26]

The 'front against environmental destruction' had strong support within Folksam. Throughout 1968, and until the summer of 1969,

---

23 Letter from Hans Palmstierna to Sören Gunnarsson, 14 January 1968, 452/3/3 (HP ARBARK); Bria Ekwall, 'Lärare matematikutbildas: Jättekurs i radio och TV', *DN*, 16 January 1968.
24 Barbro Soller, 'Bister kampanjupptakt: Avfallsproblem olösta när industri startar', *DN*, 31 January 1968; Anon., 'Front mot miljöförstöringen i ny ungdomsgiv', pp. 4–5.
25 Anon., 'Front mot miljöförstöringen i ny ungdomsgiv', p. 5.
26 Barbro Soller, 'Bister kampanjupptakt'.

the campaign and related activities received a lot of space in the insurance company's magazine. After the launch on 30 January, it was documented that 'practically all newspapers of any importance' in Sweden had reported on the youth campaign's design and aims. The same was true of radio and television.[27] The driving force in Folksam was the secretary of its youth council, Anders Ericsson. In the spring of 1968, he travelled throughout Sweden to encourage associations and schools to participate in the campaign. In total, about thirty information and networking conferences were held – events at which Ericsson, and sometimes also Palmstierna, described the format and content in some detail. Ericsson believed that the response was greater than any previous youth programme had elicited. Throughout the country, he said, there was 'a deep commitment to and responsibility for environmental issues', and he described the situation prior to the campaign's autumn launch as 'the best imaginable'.[28]

Palmstierna's correspondence from this time contains only a few letters from primary- and upper-secondary-school pupils.[29] In contrast, many university students did write to him. One of them was Lars Emmelin. He had studied zoology, zoophysiology, and genetics in Lund and worked as an assistant at the Department of Zoology. Unfortunately, he had not been able to attend Palmstierna's lecture in Lund, which was why he was writing a letter. Had the assistant professor indeed called for 'volunteers for some kind of action'? Was he referring to Folksam's campaign? Emmelin offered to sign up without reservation for whatever it was, listing his qualifications, in order to 'facilitate the assessment of whether I can be used for anything'. In addition to his studies in natural science, he cited his many activities within the student union, from which he had gained committee experience. He particularly wanted to do 'teaching and some kind of journalistic activity'.[30]

Palmstierna quickly replied that he would forward Emmelin's letter to Anders Ericsson at Folksam. He added that the popular

---

27 Anon., 'Press och miljö', *Folksam*, 1 (1968), p. 28.
28 Anon., 'Ungdomsorganisationer och skolor "heltända" på miljövård', *Folksam*, 3 (1968), p. 39; Anon., 'Front mot miljöförstöringen', *Folksam*, 3 (1968), pp. 53–55.
29 Letter from Larseric Stoltz to Hans Palmstierna, 3 April 1968, 452/3/3 (HP ARBARK); Letter from Thomas Hedner to Hans Palmstierna, 27 April 1968, 452/3/3 (HP ARBARK).
30 Letter from Lars Emmelin to Hans Palmstierna, 5 March 1968, 452/3/3 (HP ARBARK).

movements and their various educational institutions would show a slide presentation plus a short speech by himself to the young people who were active in associations. The hope was to engage a greater number of interested younger people, who would then be given access to more study materials whereupon they would start getting ready for public hearings. Palmstierna stressed that, during both the study activities and the hearings, it was important that there were 'knowledgeable people in the background, so that the young people would not be clobbered with the usual dodges and tricks'. He hoped that Emmelin might consider taking on such a role, adding that Emmelin was welcome to visit him at the National Environment Protection Board.[31]

Whether Lars Emmelin did become involved in 'A front against environmental destruction' is unclear. A subsequent letter makes it clear that Folksam did not contact him, at least not immediately. However, the friendly relationship with Palmstierna had been established. Before Palmstierna paid a new visit to Lund in May 1968, Emmelin wondered if he and his fiancée might offer 'lunch, dinner, supper' or the like. He said that he had the run of his parents' house and offered Palmstierna 'accommodation and a work space', a friendly welcome and breakfast.[32] There is no doubt that Emmelin was keen to strengthen the contact and initiate new shared ventures. In this he became successful. In the autumn of 1968, together with some other students and part-time instructors – and with Palmstierna's direct support – he organized Sweden's first course in environmental protection.

Key groups and student involvement

Hans Palmstierna and Folksam regarded young people as drivers of long-term social change. Many young people shared this view. In Palmstierna's correspondence, this mindset is manifested in a variety of ways. One example is when Wolter Arnberg, editor of *Fältbiologen* [The field biologist], contacted him with a request for an article in January 1968. Arnberg stressed that the magazine's young readers were very knowledgeable and could be expected to be 'tomorrow's opinion-makers on the environmental

---

31 Letter from Hans Palmstierna to Lars Emmelin, 7 March 1968, 452/3/3 (HP ARBARK).
32 Letter from Lars Emmelin to Hans Palmstierna, 1 May 1968, 452/3/3 (HP ARBARK).

front'.³³ Palmstierna was not hard to persuade and soon submitted a script.

Håkan Sundberg, a student at Chalmers University of Technology in Gothenburg, expressed himself even more precisely. He said that 'Chalmerists' were a conservative group of students 'who, ridiculously enough, gain access to strategic positions in the expansion society merely by virtue of their education'. It was therefore crucial that they be 'given a jolt'. Sundberg wondered whether Palmstierna might consider writing an article on the theme of 'the engineer's responsibility' for the student magazine, *Tofsen*. He also asked if Palmstierna could come to Gothenburg in March to participate in a podium debate.³⁴ Palmstierna's reply took almost a month, for which he apologized, citing his new position at the National Environment Protection Board. For the same reason, too, he did not have the opportunity to 'prod your conservative lobsters so early this year'. However, he was anxious to come and speak to the Chalmerists and hence asked Sundberg to contact him again at some point. Palmstierna also encouraged Sundberg to take an active part in Folksam's campaign and passed on his contact details to Anders Ericsson.³⁵ A month or so later, Palmstierna submitted the article 'Teknik i livets tjänst' [Technology in the service of life].³⁶

In February, Palmstierna was also contacted by Lennart Lindqvist, a student at the Agricultural College of Sweden in Uppsala. A representative of the student association JUF (Jordbrukare Ungdomens Förbund [Agricultural Youth Association]), Lindqvist wondered if Palmstierna could come and speak to them one evening in April. Lindqvist stressed that all the students at the college came into contact with 'the problems connected with the destruction of the environment and of natural resources' and that these issues directly affected their 'future professional practice'. Consequently, the students were very keen to '[have] the problems clarified as comprehensively

---

33 Letter from Wolter Arnberg to Hans Palmstierna, 18 January 1968, 452/3/3 (HP ARBARK).
34 Letter from Håkan Sundberg to Hans Palmstierna, 2 February 1968, 452/3/3 (HP ARBARK).
35 Letter from Hans Palmstierna to Håkan Sundberg, 25 February 1968, 452/3/3 (HP ARBARK).
36 Hans Palmstierna, 'Teknik i livets tjänst', *Tofsen* 4 (1968), 10–14.

as possible'.³⁷ Palmstierna accepted the invitation, saying that he was grateful for the opportunity to reach out to this 'key group'.³⁸

The most ambitious student initiative in the spring of 1968, however, occurred in Lund. A high-profile lecture series was to be held there with a focus on world poverty, overpopulation, and the difficult situation of developing countries. The keynote speaker was John Kenneth Galbraith, the world-renowned public debater and professor of economics at Harvard.³⁹ In addition, the Lund students managed to invite Raúl Prebisch, the secretary-general of the United Nations Conference on Trade and Development (UNCTAD). They also booked in prominent Swedish figures, such as Professors Gunnar Myrdal and Georg Borgström, plus the influential social democratic theorist Gunnar Adler-Karlsson. The students were eager for Hans Palmstierna to participate as well. They stressed that the event was 'completely apolitical', with the exception of a planned public opinion meeting on about 20 March which aimed to put pressure on the Swedish government. The students intended to demand that Sweden 'immediately meet the developing countries' demands at UNCTAD II'. The organizers expected that Palmstierna would 'attract a full house (1,000 people)', which was twice as many as at a regular 'student evening'.⁴⁰ Thanks to Palmstierna's participation, the Lund campaign came to encompass environmental issues; but they were hardly a primary consideration.

Even so, environmental involvement in the Scanian student city was growing. By May, Palmstierna was already there again. At that time, the law students' association, Juridiska föreningen, held a debate in the Academic Society's large auditorium on the theme of 'Environmental destruction – The price of prosperity?' The students underlined that environmental destruction was 'a most acute problem' of 'crucial importance to the existence of future generations'. What

---

37 Letter from Lennart Lindqvist to Hans Palmstierna, 18 February 1968, 452/3/3 (HP ARBARK).
38 Letter from Hans Palmstierna to Lennart Lindqvist, 25 February 1968, 452/3/3 (HP ARBARK).
39 Rolf Lindblad, 'Galbraith i Lund', *DN*, 17 March 1968. For a study of Galbraith's significance in Scandinavian politics and social debate, see Björn Lundberg, 'The Galbraithian Moment: Affluence and Critique of Growth in Scandinavia, 1958–1972', in Östling, Olsen, and Larsson Heidenblad (eds), *Histories of Knowledge in Postwar Scandinavia*, pp. 93–110.
40 Letter from Sven Herner to Hans Palmstierna, January 1968, 452/3/3 (HP ARBARK); Letter from Sven Herner to Hans Palmstierna, 26 February 1968, 452/3/3 (HP ARBARK).

could be done to stop it? What role did legislation play? In addition to Palmstierna, the other participants included Christer Leijonhufvud of the Swedish National Federation of Industry and a senior judge, Ingemar Ulveson.[41]

The following week, Palmstierna travelled to Gothenburg to officially open the provocative exhibition *Än sen då?* [So what?]. It was the brainchild of a group of architecture students at Chalmers, including the previously mentioned Håkan Sundberg. The exhibition was built in room-size sections of corrugated cardboard. Each room confronted the visitor with texts and pictures portraying humanity's living conditions. One section was about the Earth's limited resources. Another dealt with the rich world's luxury consumption and space travel. A third highlighted developing countries' poverty and lack of contraception. In the background, a counting device ticked softly at three beats per second. *Göteborgs-Posten*'s reporter informed readers that in just six hours the counter had reached 66,557. That was how many people had been born into the world since the opening of the exhibition.[42]

Än sen då? made a powerful impact. It was exhibited for nine days in Gothenburg, whereupon it set out on a national tour. In June it arrived in Stockholm, where it was opened by the then-minister for education, Olof Palme. *Dagens Nyheter* described it as 'an unusually intelligent way' of uttering a protest.[43] *Svenska Dagbladet* took up a similar position, stressing that the exhibition did not restrict itself to addressing the problems of developing countries' problems and global starvation; '[e]nvironmental destruction in all its fantastic forms is illuminated effectively and powerfully'. For example, the exhibition showed that two metro lines transported as many people as sixty car lanes. 'Which alternative is the most economical, the most socially beneficial?' asked the writer. 'Which alternative makes the greatest contribution to the stemming of environmental destruction?'[44]

---

41 Letter from Gunilla Hasselmo to Hans Palmstierna, 30 April 1968, plus the enclosed cutting from the student newspaper *Iuset*, 452/3/3 (HP ARBARK).
42 C. A., 'Medan vi väntar på världskatastrofen: Se på debattutställningen "Än sen då?"', *GP*, 15 May 1968.
43 Viveka Vogel, 'Skärpt sätt att protestera', *DN*, 13 June 1968.
44 Mari, 'Proteinrikt fiskmjöl blir kattmat i USA', *SvD*, 13 June 1968.

Schools and the environment

Hans Palmstierna was keen to spread his message in primary and upper secondary schools. Through Folksam's campaign plus recordings of environmental programmes for Skol-TV, he sought to involve teachers and pupils. By the spring term of 1968, however, he still had not produced any educational material. Teachers who wanted to introduce environmental issues into their tuition therefore had to strike out on their own. Traces of their initiatives may be found in correspondence and press clippings.

At the beginning of March an upper-secondary-school teacher named Sven-Åke Kroon wrote to Palmstierna, saying that he had read *Plundring, svält, förgiftning* several times and that the book had made him 'very frightened'. Kroon taught the subject of energy to future graduates of the school's engineering programme. He explained that the subject was closely tied to environmental issues, but that these aspects were not included in the curriculum. 'I am totally convinced that this state of affairs has to change', he said, adding that he did not intend to wait for others to act. Fortbildningsinstitutet in Stockholm [a further-education institution, later incorporated into the Stockholm Institute of Education, now part of Stockholm University – *translator's note*] had commissioned him to teach a further-education course for upper-secondary-school teachers on the subject of energy. The course would take place in the ski resort of Åre in August, and Kroon planned to focus on environmental issues. He wondered if Palmstierna could give a guest lecture, which Kroon hoped would act as a wake-up call for the energy teachers.[45]

Palmstierna immediately answered that he was interested, but that there was one drawback: recordings for Skol-TV were scheduled for the late summer. 'If this does not conflict with your course, naturally I will come.' He added that it was appalling that the upper-secondary-school engineers' curriculum did not include biology, let alone ecology. 'These, after all, are the people who must subsequently make decisions that are of the utmost importance for our environment, decisions that are now being made without any knowledge base.' He hoped Kroon and his colleagues would point out these shortcomings via their trade union.[46]

---

45 Letter from Sven-Åke Kroon to Hans Palmstierna, 6 March 1968, 452/3/3 (HP ARBARK).
46 Letter from Hans Palmstierna to Sven-Åke Kroon, 7 March 1968, 452/3/3 (HP ARBARK).

Another initiative was taken in the spring of 1968 by pupils at the technological upper-secondary school in Luleå. In collaboration with the local healthcare agency, they planned a joint action 'to clean up the city of Luleå and its surroundings'. To be launched in mid-May, the campaign would urge the general public 'to take to the streets, one and all'. The next day, the technology pupils would 'add the finishing touches and clean up the areas that had been missed'. In addition to the cleaning initiative, the 'Clean Luleå' campaign included prize competitions, guest lectures, and a podium debate. The upper-secondary-school pupils were eager for Palmstierna to attend, but he was unable to do so.[47] A few days later, he also declined a request from a senior upper-secondary-school teacher, Gunnar Ander, to come and speak in front of an 'informal group of academics' in Bromma.[48]

The biggest school venture in the spring of 1968 occurred at a primary school in Gothenburg. In the second half of April, the school intensified its environmental instruction. The aim was 'to awaken pupils' realization of the importance of environmental and nature conservation measures for tomorrow's society'. The core instruction was provided in the form of two exhibitions – one designed by the school and the other by the Society for the Promotion of Ski Sport and Open Air Life in Sweden. The arrangement began on 16 April with a lecture by the county's nature conservation officer and ended on 29 April with a lecture by Hans Palmstierna. A podium debate followed, including Palmstierna, local politicians, and officials with relevant responsibilities. To 'stimulate the pupils' own involvement', essay, drawing and photo competitions were organized on the theme of the environment.[49]

On his own initiative, a teacher named Lennart Rådström prepared a mimeographed outline entitled 'Humans in the biological environment'. Five pages long, it began with a quotation from *Plundring, svält, förgiftning*. Rådström emphasized that rapid technological development and explosive population growth had disrupted nature's state of equilibrium. Humanity was now on the brink of destroying its own habitat. For that reason, active environmental protection

---

47 Letter from Larseric Stoltz to Hans Palmstierna, 3 April 1968, 452/3/3 (HP ARBARK).
48 Letter from Gunnar Ander to Hans Palmstierna, 8 April 1968, 452/3/3 (HP ARBARK).
49 Anon., 'Miljöförstöringsproblem nytt ämne i Partilleskola', *GP*, 10 April 1968.

appeared to be a basic prerequisite for humanity's 'continued existence on Earth'. The pupils were tasked with 'drawing some simple food chains' plus a picture of how plants and fish in an aquarium were co-dependent.[50]

Porthälla School's local initiative was matched by earnest discussions at the national level. On 30 May, the National Board of Education held a conference about teaching environmental conservation. Barbro Soller reported that the conference clearly showed that 'the school system's current effort is inadequate, to say the least'. She was particularly concerned that the upper-secondary-school pupils in the engineering and economics programmes had no opportunity to receive instruction in the subject. How, then, would they be able to 'find the practical solutions for tomorrow's environmental protection'? The Board's consultant on school issues, Stig Fred, emphasized that the curriculum had been drawn up the early 1960s and that time had 'partially passed it by'. Sven-Anders Björsne, who worked at the teacher-training college in Malmö, felt that the National Board of Education should intervene with full force as soon as possible. The urgent problem called for special treatment in the same way as had previously happened with regard to sex education.[51]

There was widespread agreement at the conference that the school curriculum neglected environmental issues. The National Board of Education announced that it had set up an expert group who had been told to ensure that the issues were given more space in the next curriculum. It was also stated at the conference that school pupils themselves had begun to react. Lennart Hultgren, a teacher and member of the Biology Teachers' Association, said that one of his pupils had expressed it as follows: 'What exactly are you giving us young people with your best-welfare attitude? Well, bad air, contaminated water, toxic food – a real shit society!' The comment was quoted in both *Svenska Dagbladet* and *Dagens Nyheter*.[52]

---

50 Lennart Rådström, *Människan i den biologiska miljön*, 1968, 452/2/1 (HP ARBARK).
51 Barbro Soller, 'Läroplanen lider av fläcktyfus. Torftig miljövårdsundervisning', *DN*, 31 May 1968.
52 *Ibid.*; Monique, 'Läroplan och lärarutbildning anpassas till miljövårdsfostran', *SvD*, 31 May 1968.

The various forms of environmental involvement

The adult world could hardly be said to be indifferent to the environmental problems, though. Many people had woken up by 1968. This trend may be illustrated with figures from the National Library of Sweden's digitalized archive of the Swedish daily press. For example, we find 120 hits for the word 'miljövård' [environmental protection] for the year 1967 and 891 hits for the year 1968.[53] Research done at that time confirmed this trend. In the summer of 1968, professor of pedagogy Åke W. Edfeldt conducted telephone interviews which showed that the majority of the general public had good knowledge about mercury poisoning in nature and considered the media reporting on the issues to be reasonable. It was also possible to prove that fish consumption had fallen in connection with the publication of various alarming reports.[54] At the Riksdag level, twenty-one motions were presented that year on the topic of protecting nature and the environment, to be compared with a mere seven the year before.[55]

However, these types of quantitative measures are blunt tools for investigating the breakthrough of environmental issues in Sweden. Both linguistic usage and the understanding of the environmental problems were in a state of rapid flux. This makes it difficult to know what to compare with what in order to map lines of development over time. Concepts such as 'nature conservation' and 'environmental protection' were used interchangeably, and they could – but did not have to – refer to the same phenomenon. A specific term might have certain connotations in 1966 and quite different ones in 1968. For example, the dangers of mercury-contaminated watercourses and acid rain were sometimes linked to a global set of problems and sometimes not. Given these reservations, though, the above-mentioned quantitative readings of the national pulse do indicate that there was a degree of intensification in 1968. This is also consistent with the ways in which the historical actors themselves, such as Birgitta Odén and Sören Gunnarsson, commented on the phenomenon. The change is palpable in Hans Palmstierna's correspondence, too. By scrutinizing that

---

53 Search on the word 'miljövård' at www.tidningar.kb.se (accessed 29 May 2020).
54 Åke W. Edfeldt, *Kvicksilvergäddan* (Stockholm: Tiden, 1969).
55 Record of proceedings in the Riksdag plus appendices 1961–1970. Vol. 3, Index, L–Ö.

development, we can move close to the historical process of change in a qualitative way.

One of the people who began to take an interest in environmental issues in the spring of 1968 was Eric Bergh in Gothenburg. He introduced himself to Hans Palmstierna as 'one of the many Gothenburgers who do not usually demonstrate or otherwise make our voice heard'. Bergh expressed his 'heartfelt thanks' to Palmstierna for the latter's great efforts in the field of environmental protection and described how he and some friends had compiled and circulated a petition. By such means, the group sought 'to express directly what those of us at the popular level want in these matters'. They had hence been careful to ensure that only people 'who had read it and had taken a firm personal stand' signed the petition. The group had not allowed any representatives of associations or organizations to sign collectively. In just a few days, 700 signatures had been collected. The group's petition had been noticed and cited in the press, and it had begun to be circulated within the city administration. Bergh enclosed a copy of the petition in his letter to Palmstierna, so that the latter could gain insight into what was happening in Gothenburg prior to his upcoming visit there.

The first sentence of the petition emphasizes that it was precisely the adult world that was raising its voice. 'We, families in Gothenburg, protest against our children's having to grow up and develop in an *increasingly poisoned environment*, created by our senseless scurrying after material standards and our mentality of earnings and profit.' The wording of that protest contained echoes of Hans Palmstierna's own rhetoric. The signatories of the petition demanded that measures be implemented immediately. They wanted to see a ban on leaded petrol, on the mass emission of biocides, and on the construction of new multistorey car parks in the city centre. They called for mandatory exhaust filters on cars and buses, better purification systems in factories, and the construction of new sewage-treatment plants. They also wanted to see more effective measures against 'alcohol and drug abuse, a ban on advertising for alcohol and tobacco, [and] strong measures for the cure and rehabilitation of alcoholics'. The detailed passage about abuse may seem to be a poor fit in the context. After all, the appeal's focus was on fresh air, clean water, and a healthy living environment.[56] But it indicates a general tendency

---

56 Letter from Eric Bergh to Hans Palmstierna, 15 May 1968, 452/3/3 (HP ARBARK).

in the material. The commitment to the environment was not an isolated phenomenon; it was linked to other issues that people cared about. For some, those issues were development at home and abroad, overpopulation, and world peace; for others, they were experiences of nature, historical research, and alcohol abuse.

People with artistic ambitions were able to manifest their environmental commitment via forms of aesthetic expression. One such individual was the troubadour Anders Fugelstad. In June 1968, he sent some printed song lyrics to Palmstierna. Fugelstad stressed that he was happy for them to be used and that he himself would be available for suitable assignments. His lyrics contained lines such as 'the population is increasing every minute / the love of raw materials must end' and 'I suppose our prosperity must cost us our well-being / and so far I still have the strength to cough'. He sang about horses that had died of carbon-monoxide poisoning and mountain streams that had become polluted rivers.[57] The same approach was adopted by the established poet Stig Carlsson, who dedicated the poem 'Om vissa förutsättningar' [About certain conditions] to Hans Palmstierna. The poem levelled harsh criticism at space travel and global injustices. Carlsson argued that humanity should focus its attention on the face of the Earth rather than on the far side of the moon. The most important thing was good harvests and that people could eat their fill. 'And this strikes me', he concluded, 'as an almost childish opinion.'[58]

The summer of 1968 also saw examples of organized student involvement. The most ambitious case was a two-week seminar event at the Konstfack School of the Applied Arts in Stockholm [now Konstfack University of Arts, Crafts and Design] on the theme 'Humanity – the Environment'. The guest lecturers included Carl-Göran Hedén and Tor Ragnar Gerholm. The students devoted themselves to workshops and group projects. Prior to the event, they edited a special issue of the magazine called *Form*, which featured the responsibility of design students for the development of society.[59] Palmstierna was invited to speak, but could not attend. However, in a radio talk entitled 'De klarsyntas revolt' [The rebellion of the clearsighted ones], he expressed his great support for students

---

57 Letter from Anders Fugelstad to Hans Palmstierna, 14 June 1968, 452/3/3 (HP ARBARK).
58 Stig Carlsson, *Förbifarter* (Stockholm: Norstedts, 1968), pp. 38–39.
59 Letter from Eva Lamby to Hans Palmstierna, June 1968; Rebecka Tarschys, 'Konstfackare vill ta ansvar', *DN*, 22 July 1968.

around the world who turned against established institutions in society in the late spring of 1968. 'The young people are rebelling because they want to continue living', he pointed out. The students were rightly attacking those who were steering the world 'straight into a coming hell'. It was not defensible to send in 'vigilantes and riot police' against this uprising. On the contrary, society should invite the young generation to 'plan the society in which they will be living'. That was a prerequisite for building a better world and ensuring humanity's survival.[60]

The environmental issues intersected with the lives of older people as well. One of them was Valfrid Irskogen from Malmö. He had been told by a relative that the mining company Boliden 'cast arsenic into cement blocks' and then dumped them in the Gulf of Bothnia. This behaviour, if true, frightened him. In his childhood in the 1890s, many people had used 'arsenic dissolved in water on a plate as fly poison'. For that reason, he contacted Palmstierna to inform him of his suspicions.[61] Another elderly man from Skåne who wrote to Palmstierna was Zenon P. Westrup. A retired diplomat and former chairman of the local nature-conservation association in Malmö, Westrup introduced himself as 'an old labourer in the vineyard' who had great interest in and sympathy for 'your teachings in the context of nature conservation'. However, Westrup was disappointed that Palmstierna was mixing politics into the issues. That was, he wrote, 'the most disastrous thing that could happen to our common interests in this field'. He especially objected to Palmstierna pinning his hopes on the Social Democratic movement. To Westrup, such a view was almost grotesque. As a 'reasonably good right-winger' he had for decades 'fought, written, quarrelled and become enemies with people' by pushing nature-conservation issues at the local level. His opponents had most often been Social Democrats. Still, Westrup did not want to single out that party as being worse than others. Instead, he argued that 'the entire Swedish people' had constructed 'the same simple and barbaric scale of values with regard to priorities between material benefits and environmental values' over a long period of time.[62]

---

60 Hans Palmstierna, 'De klarsyntas revolt', in *Angeläget idag* (Stockholm: Sveriges Radio, 1968), pp. 128–129.
61 Letter from Valfrid Irskogen to Hans Palmstierna, 7 July 1968, 452/3/3 (HP ARBARK).
62 Letter from Zenon P. Westrup to Hans Palmstierna, 18 July 1968, 452/3/3 (HP ARBARK).

Palmstierna responded by sending *Plundring, svält, förgiftning* to Westrup. It was received as 'an exquisite and disarming expression of amiability'. Westrup wrote that he was ashamed of not having read the book when it had first come out, adding that it was Birgitta Odén who had drawn his attention to Palmstierna's activities. However, he still insisted that the politicization of nature-conservation issues posed a danger to their shared interests. Socialist politicians, just like progress-seeking business leaders, were easily seized 'by the same idolization of technology' and by economistic perspectives. A look at 'the former and the current head' of the state-owned energy company Vattenfall was proof enough, he said.[63]

Westrup's two letters to Palmstierna are of general interest. They provide insights into the conflict-filled encounter between older nature-conservation interests and the emerging policy of environmental protection. The former had roots in the late nineteenth century when (among other things) the Swedish Society for Nature Conservation was formed. Its focus lay on protecting the wilderness from industrial expansion, for example by founding national parks. Nature conservation had traditionally had a bourgeois flavour with close ties to the university world,[64] and this background is obvious in Westrup's letters. His path to Palmstierna did not go via the press, the book market, or popular movements. It went via Professor Birgitta Odén, whom he had contacted after she had sent him the programme for 'some kind of "study group" for environmental issues'.[65]

Outside Sweden's borders

The intertwined breakthroughs for environmental issues and for Hans Palmstierna personally were a Swedish phenomenon. In most other countries, a commitment to the environment was still a marginal phenomenon in 1967 and 1968. The major exception was the United States, where Barry Commoner and Paul Ehrlich attracted

---

63 Letter from Zenon P. Westrup to Hans Palmstierna, 1 August 1968, 452/3/3 (HP ARBARK).
64 Désirée Haraldsson, *Skydda vår natur!: Svenska naturskyddsföreningens framväxt och tidiga utveckling* (Lund: Lund University Press, 1987); Jonas Anshelm, *Det vilda, det vackra och det ekologiskt hållbara: Om opinionsbildningen i Svenska Naturskyddsföreningens tidskrift Sveriges natur 1943–2002* (Umeå: Umeå universitet, 2004).
65 Letter from Zenon P. Westrup to Hans Palmstierna, 18 July 1968, 452/3/3 (HP ARBARK).

considerable attention in the public debate during those two years.⁶⁶ However, in contrast to Palmstierna, they were far from the centre of political power. In a letter to Olof Palme, Palmstierna said that his scientific friends in the United States 'envy us our opportunities to reach out to people with a message via the popular movements'. Besides, he added, the American mass media were sitting on the lap of industrial and commercial interests. As a result, they could not 'spread knowledge about such matters as the destruction of nature and resources in the same way as we [can] here in Sweden'.⁶⁷

Against this background, it is hardly surprising that the lion's share of Palmstierna's surviving correspondence is in Swedish. However, some threads do lead outwards. In January 1968 he was contacted by the Dutch lawyer Jacques de Reus, who had previously lived in Stockholm. De Reus had become aware of Palmstierna through *Dagens Nyheter*'s series of articles in the 'environment of the future'. The Dutchman said that environmental problems 'apply even more' to a densely populated, industrialized country such as the Netherlands. He therefore wondered whether Palmstierna might be interested in a translation into Dutch. He himself would happily 'act as a middleman to persuade a publisher of the book's importance'.⁶⁸ Palmstierna replied enthusiastically – in Dutch. He and his sister had spent part of their childhood in the Netherlands during the Second World War, and de Reus's letter became the start of a lengthy correspondence.⁶⁹

Still, most of non-Swedish interest in Palmstierna's operations came from the neighbouring Nordic countries, where many people were following the Swedish developments. One of them was the Finnish-Swedish student Richard Ahlqvist. He was studying architecture at the Helsinki University of Technology and planned to devote his graduation-essay project to waste management. For him,

---

66 Adam Rome, '"Give Earth a Chance": The Environmental Movement and the Sixties', *Journal of American History* 90.2 (2003); Egan, *Barry Commoner and the Science of Survival*; Robertson, *The Malthusian Moment*; Rome, *The Genius of Earth Day*.
67 Letter from Hans Palmstierna to Olof Palme, 27 April 1968, 452/3/3 (HP ARBARK).
68 Letter from Jacques de Reus to Hans Palmstierna, 17 January 1968, 452/3/3 (HP ARBARK).
69 Letter from Hans Palmstierna to Jacques de Reus, 23 January 1968, 452/3/3 (HP ARBARK). For his childhood, see Gunilla Palmstierna-Weiss, *Minnets spelplats* (Stockholm: Bonnier, 2013).

the whole topic had 'grown much larger than just an interest'. He explained that his entry point had been Palmstierna's article in *Dagens Nyheter* entitled 'Förskingringen kan hejdas' [The embezzlement can be stopped] (21 March 1967). Consequently, he wanted to meet Palmstierna to discuss his ideas and said he could come to Stockholm at any time.[70]

The reply has not been preserved, but from Ahlqvist's next letter it is clear that Palmstierna had suggested meeting in Helsinki on 6 June. Ahlqvist could not make that date, though. He added, in a tone of disappointment, that his planned degree project had run into opposition. The Department of Architecture's faculty felt it was too extensive. Besides, they thought that Ahlqvist lacked the proper qualifications for the job and suggested that he engage in 'something more artistic'. Ahlqvist had protested in vain, and his plans looked like remaining at the idea stage. He nevertheless thanked Palmstierna for the latter's kind letters and stressed that environmental problems were still 'unresearched areas in Finland'. He therefore hoped he could 'do his own small part to add to a growing awareness of and opposition to the mismanagement'. He was also pleased to see that Palmstierna 'had been contacted about a major summer seminar at Sveaborg in July'. Perhaps they might get together there?[71]

The impression that Sweden was at the forefront of environmental protection was also expressed by a diplomat working in Oslo, Ivar Öhman. He said that Norway had not yet 'arrived at our "tough" views and demands'. He was contacting Palmstierna because he had seen the exhibition *Än sen då?* during a visit to Stockholm. He had found it 'damned disturbing and important' and therefore wanted to bring it to Oslo, where he felt it could 'do a lot of good'. He envisioned lectures, podium debates, and film screenings in conjunction with the exhibition. 'Of course' the event would 'target a young audience'. Might Palmstierna have the contact details of the students behind the exhibition?[72]

Another Swede living abroad who wanted to spread Palmstierna's message was H. William-Olsson. He lived in London, where he was the honorary secretary of the Anglo-Swedish Society. The Society

---

70 Letter from Richard Ahlqvist to Hans Palmstierna, 4 May 1968, 452/3/3 (HP ARBARK).
71 Letter from Richard Ahlqvist to Hans Palmstierna, 30 May 1968, 452/3/3 (HP ARBARK).
72 Letter from Ivar Öhman to Hans Palmstierna, 3 July 1968, 452/3/3 (HP ARBARK).

sought 'to act as a catalyst in the Swedish/English exchange of ideas'. William-Olsson wondered if Palmstierna was planning to visit London in the autumn. Unfortunately, the Society did not have the funds to pay for the trip, but its members were eager for him to come and speak to them. William-Olsson explained that they had good contacts with the press and the BBC, and that the lecture could therefore be a springboard for Palmstierna to reach a wide audience. He was also sure that there was great interest in London in Sweden's National Environment Protection Board, this 'totally unique initiative'.[73]

Palmstierna was interested and suggested that an event could be held in conjunction with his trip to the Netherlands towards the end of the year. William-Olsson quickly replied and again regretted the Society's poor finances. He stressed that an appearance 'would be such a definite Swedish interest' that it should be possible to obtain funding. Perhaps from the National Environment Protection Board or the embassy? He added that it was not '*completely* unthinkable' that some form of appearance on the BBC might result. He could not promise anything, but intended to invite 'selected members of the press'.[74]

The type of chain reaction that William-Olsson hoped for in London occurred in Finland in August. Following an appearance in Helsinki, Palmstierna was contacted by the television editor Lauri Markos. He explained that Palmstierna's lecture had made a strong impression on him and that a programme was now being planned for Finnish television about the threats to the human habitat. Markos was particularly interested in the dangers that 'can be found in various foods' and wondered if Palmstierna could help to turn the programme idea into reality.[75] A few weeks later, Teuvo Suominen of the Finnish Nature Conservation Society wrote to Palmstierna. Together with Vaasa Summer University, the Society planned 'to organize a week for environmental protection' in the summer of 1969. He would like to invite Palmstierna as a speaker, and he also wondered if it might be possible to bring in the exhibition *Än sen då?* In conclusion, he said that *Plundring, svält, förgiftning* had attracted great attention in Finland and he hoped it would soon be

---

73 Letter from H. William-Olsson to Hans Palmstierna, 5 August 1968, 452/3/3 (HP ARBARK).
74 Letter from H. William-Olsson to Hans Palmstierna, 25 August 1968, 452/3/3 (HP ARBARK).
75 Letter from Lauri Markos to Hans Palmstierna, 26 August 1968, 452/3/3 (HP ARBARK).

translated. 'We are very impressed here by the new generation of Swedish prophets who have arisen, individuals like Rolf Edberg, Nils Landell, Gunnar Myrdal, yourself, etc.'[76]

In Denmark, too, people were starting to become aware of Hans Palmstierna. In the early autumn of 1968, his book was favourably reviewed in the Copenhagen-based morning newspaper *Politiken*. Its editor, Harald Mogensen, wondered if Palmstierna himself would be interested in writing a column for the newspaper.[77] Palmstierna immediately said yes.[78] Mogensen explained that in Denmark, they 'had not really had a comprehensive debate about the immense scope and significance of the pollution problems'. Occasionally there were small discussions, but 'the big perspective has, as it were, not been fully outlined'. He hoped that a column by Palmstierna could change that.[79] In addition to Mogensen, the youth wing of the Danish Social Democratic Party also contacted Palmstierna to invite him to speak to them. He told them about 'the educational activity that exists in the field of environmental protection in Sweden'. He singled out Folksam's campaign and encouraged the Danish young people to contact Anders Ericsson.[80] Nordic interest in the campaign grew to a considerable level. At the beginning of 1969, conferences were held in Copenhagen and Oslo at which Hans Palmstierna and Folksam spoke about the Swedish initiative. The Folksam campaign became a model for similar campaigns in Denmark, Norway, and Finland.[81]

The environmental awakening in Sweden

The greatest interest in environmental issues, as well as in Hans Palmstierna, was found in Sweden, however. In August 1968, an anonymous light-hearted article in *Dagens Nyheter* portrayed him as 'the revivalist preacher'. The writer described how jealousy was

---

76 Letter from Teuvo Suominen to Hans Palmstierna, September 1968, 452/3/3 (HP ARBARK).
77 Letter from Harald Mogensen to Hans Palmstierna, 27 September 1968, 452/3/3 (HP ARBARK).
78 Letter from Hans Palmstierna to Harald Mogensen, 30 September 1968, 452/3/3 (HP ARBARK).
79 Letter from Harald Mogensen to Hans Palmstierna, 7 October 1968, 452/3/3 (HP ARBARK).
80 Letter from Hans Palmstierna to Görgen Christiansen, 11 October 1968, 452/3/3 (HP ARBARK).
81 Anon., 'Nordisk front mot miljöförstöring', *Folksam*, 4 (1969), p. 31.

sweeping the scientific world and fury was now simmering among the industrial and municipal big shots: 'And all this while almost all the rest of the Swedish people just want to have more Palmstierna.' He had 'offers for far more than 365 speeches a year' and was constantly on the move. But why 'this tidal wave of Palmstierna through all our media?' The writer of the article explained that *Plundring, svält, förgiftning* had appeared at just the right time. Palmstierna himself had a 'gentle voice' and managed to be 'placid and amiable' even though he 'preached terrible truths'. In addition, he was able to play 'on the entire communication apparatus'. But could he really continue to keep up this pace? Was he not out and about a little too often?[82]

The humourist's questions came to a head in the autumn of 1968. The 'front against environmental destruction' had been launched in earnest and the Social Democrats' campaign for the upcoming Riksdag election was intensifying. Palmstierna declined to stand for a seat in the Riksdag himself, but he lent his name to the campaign and spoke at rallies. In a large published advertisement, he explained that 'as a scientist with some degree of insight into humanity's problems and future, one cannot but take a political stand'. In a world of overpopulation, oppression, environmental destruction, lust for power, and commercialism, it was impossible just to stand by. The great enemy was self-interest; the answer, solidarity. 'A socialism with open dialogue and free discussion of the problems and their solutions then becomes the self-evident option. That is why I am a Social Democrat.'[83]

Letters from the public continued to pour in. Gunilla Brotaeus, a student, wrote that she was 'fully aware of what a busy person' she was bothering, but she still hoped that 'someone who so stubbornly spreads his message cannot be completely uninterested in the fish that bite'. Brotaeus wrote that she was studying English at Stockholm University and was in the process of writing an essay on environmental destruction. Unfortunately, she wrote, she did not believe that she had sufficiently comprehensive material, particularly regarding the economic aspects. She therefore wondered if Palmstierna could help her disseminate the facts 'even though it be on a small scale'.[84] Palmstierna thanked her for her interesting letter but did not send

---

82 Anon., 'Hans Palmstierna, väckelsepredikanten', *DN*, 11 August 1968.
83 Anon., 'Egennyttan den stora fienden', *DN*, 1 September 1968.
84 Letter from Gunilla Brotaeus to Hans Palmstierna, September 1968, 452/3/3 (HP ARBARK).

her any information, advising her instead to contact Professors Erik Dahmén and Birgitta Odén, plus Anders Ericsson at Folksam.[85]

That same day, Mats Börjesson in Umeå received a similar reply. He represented the city's newly formed development group and informed Palmstierna that the Museum of Västerbotten had bought the rights to show the exhibition *Än sen då?*. It would be in Umeå during the month of October and then tour the Västerbotten province. In conjunction with this, the group was planning 'to conduct an elementary information campaign in order to involve more people'. Could Palmstierna possibly write something for the newspaper *Västerbottens folkblad* and come to Umeå and speak at some point during that autumn?[86] He could not – he was fully booked all autumn. He suggested that Börjesson should contact Lennart Danielsson at the National Environment Protection Board or Associate Professor Göran Löfroth at the Department of Radiation Biology at Stockholm University. As for himself, he declined 'with sadness in my heart, because I know how committed you and the people around you are'.[87]

One of those people was upper-secondary-school teacher Kerstin Hägg. She explained that Georg Borgström had recently been to Umeå and given a lecture to teachers and pupils. While 'most people still have the terrible realities that Borgström presented to us fresh in their memory', she hoped to launch a campaign that could benefit developing countries. Hägg's idea was that all the teachers at her school would set aside 'a small part of their large salaries' for this purpose. The problem was to find 'a project that as many people as possible can accept'. She was considering international family planning organizations, but wondered whether Palmstierna had any other idea? She said he had done a lot to raise her own awareness, singling out a lecture he had given in Umeå in the autumn of 1966. It had been 'something of a "wake-up call"' for her, and she now hoped that she could 'spread my little insight further' to acquaintances and pupils. Never before had she felt that the teaching profession was so very important.[88]

---

85 Letter from Hans Palmstierna to Gunilla Brotaeus, 30 September 1968, 452/3/3 (HP ARBARK).
86 Letter from Mats Börjesson to Hans Palmstierna, 20 September 1968, 452/3/3 (HP ARBARK).
87 Letter from Hans Palmstierna to Mats Börjesson, 30 September 1968, 452/3/3 (HP ARBARK).
88 Letter from Kerstin Hägg to Hans Palmstierna, 7 September 1968, 452/3/3 (HP ARBARK).

Palmstierna was pleased by the letter and answered it at length. Stressing the importance of 'making people as aware of the problems as possible', he said that trying to 'enlighten the entire population' was not a reasonable endeavour. The way forward was 'to start by focusing on particular groups'. His own immediate goal was to cause 'a change of heart' among the people who prepared the ground for political decisions. The attitude of the decision-makers themselves was much harder to change. Palmstierna considered Folksam's campaign to be one stage of this work, and he encouraged Hägg to contact Anders Ericsson. She could thereby play a leading role in the work towards change. With regard to Hägg's fundraising proposal, he was positive but brief. In his view, the key to long-term success was primarily found in shaping opinion and changing attitudes.[89]

The growing interest in the environment found in Umeå is also evident in a letter from a man named Lars Gustafsson. He wrote to Palmstierna informing him that 200 people had helped to found the city's development group. Its activities were divided into ten action-and-study subgroups. Gustafsson himself represented the subgroup focusing on 'the destruction of nature and the environment'. He explained that his group was reaching out by giving speeches in schools and workplaces, holding demonstrations, and putting pressure on politicians. But he was also keen for the group to identify particular issues on which to focus its efforts. He himself was heavily involved in 'bringing about a ban on non-returnable glass' and therefore turned to Palmstierna with seven specific questions.[90] Once again, Palmstierna encouraged the letter-writer to contact Anders Ericsson at Folksam and sought to channel the public's involvement via the 'Front against environmental destruction'.[91]

Palmstierna himself no longer had enough time to deal with all the inquiries. At the beginning of the year, he had spoken at church services and local Rotary clubs.[92] By the autumn of 1968, however, he was saying 'no' to almost everything. The upper-secondary-school

---

89 Letter from Hans Palmstierna to Kerstin Hägg, 30 September 1968, 452/3/3 (HP ARBARK).
90 Letter from Lars Gustafsson to Hans Palmstierna, 26 September 1968, 452/3/3 (HP ARBARK).
91 Letter from Hans Palmstierna to Lars Gustafsson, 4 October 1968, 452/3/3 (HP ARBARK).
92 Letter from Vällingby Rotary Club to Hans Palmstierna, 28 December 1967, 452/3/2 (HP ARBARK); Letter from Sven Volk to Hans Palmstierna, 11 February 1968, 452/3/3 (HP ARBARK).

pupils at Östra Real School in Stockholm, the doctors at Jönköping County Hospital, the business leaders of the Halland County marketing association, and the initiators of International Week in Malmö all received the same negative answer as Umeå's development group.[93] However, Lars Emmelin in Lund did receive Palmstierna's active support. Emmelin was the driving force behind the student initiative that had led to the creation of Sweden's first course on environmental protection. The newspaper *Arbetet* reported that there had been many applications. Only 50 of the 200 applicants had been accepted. At the beginning of October, the course began with its formal lead instructor and examiner – Hans Palmstierna – coming to Lund and giving an introductory lecture.[94] In the autumn of 1968, the [trade-union movement's] Workers' Education Association (ABF) also began giving courses based on Palmstierna's book, and [the Liberal and Centre Party adult-education association] Studieförbundet Vuxenskolan launched the course 'Framtidsmiljön' [The environment of the future].[95] By this time, *Plundring, svält, förgiftning* had sold 29,000 copies and been published in a fifth edition.[96]

One year after environmental issues had made their major breakthrough in Sweden, the so-called environmental awakening was a fact. Knowledge of a global environmental crisis was not only circulating in the public sphere; it was part of many people's lives. One sign of the times was the publication on 30 September 1968 by *Dagens Nyheter* of a two-page spread entitled: 'Det här kan vi alla göra för miljövården' [This is what we can all do towards environmental protection]. The journalist, Gun Leander, suggested various ways in which consumers could make good environmental choices, such as drinking beer from recyclable bottles, washing dishes by hand, using low-octane petrol in their cars, and having composting toilets in their country cottages. Most important, though, was that 'we ourselves

---

93 Letter from Hans Palmstierna to Lennart Blomdahl, 20 September 1968, 452/3/3 (HP ARBARK); Letter from Hans Palmstierna to Ragnhild Billig, 20 September 1968, 452/3/3 (HP ARBARK); Letter from Hans Palmstierna to Jonas Ljungberg, 30 September 1968, 452/3/3 (HP ARBARK); Letter from Hans Palmstierna to Karl Eriksson, 30 September 1968, 452/3/3 (HP ARBARK).

94 Anon., '50 av 200 kom med på miljövårdskurs i Lund', *Arbt*, 2 October 1968. The origins and history of the course are described in detail in an unpublished manuscript which Lars Emmelin allowed me to read.

95 Torsten Sandberg, *Framtidsmiljön: En studieplan från studieförbundet Vuxenskolan* (Falköping: Studieförbundet Vuxenskolan, 1968).

96 Advertisement for *Plundring, svält, förgiftning*, *SvD*, 9 November 1968.

get involved and force our politicians to act'. Environmental issues were primarily the concern of citizens, not of consumers. That said, there were many opportunities 'for everyone with an environmental conscience' to work for change in daily life as well.[97] The beginning of 1969 saw the first regular opinion polls measuring the Swedish public's attitude to environmental destruction.[98]

The swarm of activities we have encountered in this chapter can hardly be characterized as an organized environmental movement. No such thing existed in Sweden or anywhere else at this time. By the beginning of the 1970s, though, the situation was different. That was when a number of new organizations of varying size and importance were founded. They included successful international networks such as Friends of the Earth and Greenpeace. But during the Swedish breakthrough phase of 1967 to 1968, environmental issues were being pursued within and via established organizations. Traditional nature-conservation associations form a particularly intriguing prism for studying this process. They possessed a long-cultivated commitment to protecting animals, nature, and the wilderness from the advance of modern civilization. What happened to that commitment when the environmental awakening occurred? Did operations change in the established organizations? Or did most things stay the same? With a view to answering those questions, the next chapter examines the youth organization Nature and Youth Sweden (*Fältbiologerna*).

---

97 Gun Leander, 'Det här kan vi alla göra för miljövården', *DN*, 30 September 1968.
98 Lennart Lundqvist, *Miljövårdsförvaltning och politisk struktur* (Uppsala: Verdandi, 1971), pp. 106–107; Bennulf, *Miljöopinionen i Sverige*, p. 56.

6
The emergence of the modern environmental movements, 1959–1972

On Sunday 16 March 1969, Nature and Youth Sweden – *Fältbiologerna* [literally The Field Biologists] in Swedish – held a nationwide demonstration against the expansion of hydroelectric power in northern Sweden. The biggest gathering took place in central Stockholm, where a couple of hundred people met in order to march from Östermalmstorg to Sergels Torg, a distance of approximately 800 metres. *Svenska Dagbladet* described it as a demonstration 'of a somewhat unusual kind', and *Dagens Nyheter* pointed out that the young people did not look like 'ordinary demonstrators'.[1] Their placards bore messages such as 'Killing nature is suicide', 'Welfare is a pristine river', and 'Your children are protesting against your short-sightedness'.[2] From the rostrum at Sergels Torg, Nature and Youth Sweden's chairman, Wolter Arnberg, sharply criticized the short-term economic interests which 'always and without exception' caused environmental interests to be crushed. He demanded that the plans to exploit the Vindel, Kaitum, and upper Lule Rivers be mothballed immediately 'so that the nuclear power plants can demonstrate that there are concrete alternatives to hydroelectric power'.[3]

The notion that nuclear power was environmentally friendly was well established in late 1960s Sweden. Ever since the 1950s, nature-conservation bodies had hoped that the new technology would put a stop to the continued exploitation of the great rivers

---

1 Anon., 'Kamp för Kaitum på Östermalmstorg', *DN*, 17 March 1969; Monica Anrep-Nordin, 'De vill rädda vår vildmark', *SvD*, 17 March 1969.
2 Anon., 'Kamp för Kaitum på Östermalmstorg'; Anrep-Nordin, 'De vill rädda vår vildmark'.
3 Anon., 'Tal framfört på Sergels Torg den 16.3 1969 vid Fältbiologernas demonstration mot vattenkraftutbyggnaden i fjällen', *Fältbiologen* (FB), 3 (1969), pp. 4–5.

Emergence of the modern environmental movements 145

of the north. By preserving untouched wilderness from the advance of civilization, these interest groups wanted to secure aesthetic natural values and opportunities for recreation and outdoor life.[4] The protest against the expansion of hydroelectric power on 16 March received a good deal of attention in the press and broadcast media, which claimed that Nature and Youth Sweden was not an organization that is usually 'associated with demonstration marches'.[5] That situation was about to change, though. In the years around 1970, Nature and Youth Sweden's focus and activities were recast. Whereas the young people had previously devoted themselves to birdwatching, nature studies, and camping activities, they became increasingly known for direct actions and far-reaching social criticism.[6] Consequently, the association became a visible and influential part of the emerging Swedish environmental movement. In parallel with this shift, its membership grew rapidly, from about 3,000 in the mid-1960s to over 10,000 in the early 1970s.[7]

The history of Nature and Youth Sweden dates back to 1947. At that time, Sveriges Fältbiologiska Ungdomsförening (SFU) was formed as a youth branch of the Swedish Society for Nature Conservation (SNF), which was the oldest and largest nature-conservation organization in the country. SNF's original ambition was to become a popular movement, but up until 1955 it never had more than 5,000 members. It was an expert organization dominated by a scientific elite and with close ties to the state and to the university world. In the 1960s, however, its membership base expanded, and by 1970 SNF had about 50,000 members.[8] By contrast, Nature and Youth Sweden was an independent association run by the young members themselves. It was open to everyone aged 7 to 25 years. Its aim was to spread and increase knowledge about nature among children and young people. The fundamental

---

4 Jonas Anshelm, *Mellan frälsning och domedag: Om kärnkraftens politiska idéhistoria i Sverige 1945–1999* (Eslöv: Symposion, 2000), pp. 100–102; Anshelm, *Det vilda, det vackra och det ekologiskt hållbara*, p. 32.
5 Anon., 'Kamp för Kaitum på Östermalmstorg'.
6 This chapter is based on Anna Kaijser and David Larsson Heidenblad, 'Young Activists in Muddy Boots: Fältbiologerna and the Ecological Turn in Sweden, 1959–1974', *Scandinavian Journal of History* 43.3 (2018).
7 Helena Klöfver, *Håll stövlarna leriga och för bofinkens talan: Naturintresse, miljömedvetenhet och livsstil inom organisationen Fältbiologerna* (Linköping: Tema V report, 1992), pp. 36–37.
8 Anshelm, *Det vilda, det vackra och det ekologiskt hållbara*, p. 8.

idea was that this was best done directly in the field and through peer education.⁹

Nature and Youth Sweden's activities were organized in the form of local clubs, which numbered about a hundred by the mid-1960s. Recruitment mainly happened through schools, where active members of the association spread the word about their activities. The clubs arranged excursions and meetings. Many local clubs published their own mimeographed membership magazines. The association's most important communication channel was the printed magazine *Fältbiologen* [The field biologist]. It was sent to all members six times a year. Through the magazine, it is possible to study Nature and Youth Sweden's ideological development and gain insights into its activities. From having been an apolitical association for nature-interested youth, Nature and Youth Sweden became a breeding ground and base for environmental activism. When and how did this change occur? What did it actually consist of? And what role did Nature and Youth Sweden play in and for the emergence of organized environmental movements in Sweden? To investigate these matters, I will begin by looking back at the state of Nature and Youth Sweden before the environmental issues had their big breakthrough.

Nature-interested young people, 1959–1966

In 1959, Nature and Youth Sweden's parent organization, the Swedish Society for Nature Conservation, celebrated its fiftieth anniversary. This prompted Lars-Erik Åse, a member of Nature and Youth Sweden, to reflect on the passage of history. He asserted that no other fifty-year period had been 'so rich in transformations and upheavals'. Sweden had developed into one of the world's richest countries. The prosperity and standard of living of the late 1950s could not have been dreamed of by the 'Swedish inhabitants of the turn of the century'. These rapid developments were gratifying in many ways; Swedes had gained 'better housing, better food, cars,

---

9 Thomas Söderqvist, *The Ecologists: From Merry Naturalists to Saviours of the Nation: A Sociologically Informed Narrative Survey of the Ecologization of Sweden 1895–1975* (Stockholm: Almqvist & Wiksell International, 1986); Jamison, Eyerman, and Cramer, *The Making of the New Environmental Consciousness*, pp. 18, 31; Klöfver, *Håll stövlarna leriga*; Helena Klöfver, *Miljömedvetenhet och livsstil bland organiserade ungdomar* (Linköping: Linköpings universitet, 1995); Wennerholm, *Framtidsskaparna*, pp. 248–278.

televisions, and more leisure time'. However, Åse was concerned that humanity had not kept up with this progress. Human beings did not seem to be able to keep pace with the advances of technology. Manifestations of this discrepancy included the increasing incidence of 'neuroses and nerve diseases' and the growth of juvenile delinquency.

Åse's overall explanation was that 'the modern human' had lost contact with nature and lacked *'interest* in and *feeling* for nature'. This loss made Nature and Youth Sweden's activities important, because that was precisely what the association was trying to give its members. It did so 'not by preaching and admonishing', but by cultivating individual members' fascination with the interaction between plants and animals. From there, Åse argued, it was a straight line to nature conservation; because people who had an 'interest in and feeling for nature' would automatically wish 'to protect and preserve endangered areas'. This desire could frequently lead to conflicts with industrial and economic interests. Åse concluded: 'Here the question becomes whether we might consider sacrificing some small part of our high standard of living' in order to preserve beautiful and atmospheric landscapes.[10]

The views expressed by Lars-Erik Åse in 1959 were typical of the older nature-conservation tradition within which Nature and Youth Sweden had been formed. The tradition was strongly coloured by Romanticism, which emphasized feelings and experiences of wild nature. Untouched nature was said to possess aesthetic, almost spiritual, values which modern cultural landscapes were thought to lack. Great emphasis was also placed on the scientific study of nature. Special topics such as ornithology, botany, entomology, and ecology were highly valued. For Nature and Youth Sweden, knowledge about nature and feelings for it were two sides of the same coin.

So why was it so important to protect wild nature? Karin Furuwidh, chair of Nature and Youth Sweden in 1960, was clear about the answer. 'Remember', she wrote, 'that a plentiful and untouched nature is the very foundation for our hobbies!'[11] This recreational leisure argument was repeatedly made in *Fältbiologen*'s columns during the early 1960s. Wild nature should be saved from the advance of modern civilization so that humans, not least young nature enthusiasts, could continue to enjoy it. This theme was often

---

10 Lars-Erik Åse, 'Fältbiologi och naturvård', *FB*, 5 (1959), p. 2.
11 Karin Furuwidh, 'SFU:aren och Naturvården', *FB*, 1 (1960), p. 3.

combined with historical references to rising prosperity, increasing leisure, and the emergence of motoring.[12] Anne von Hofsten stressed that these developments were placing ever-higher demands on nature, and that humanity's responsibility had increased as a result. She maintained that 'it is we who must train ourselves and other young people to interact with nature in the right way', because within a few years it is 'our generation' that will lead the country.[13]

Members of Nature and Youth Sweden were not outspoken critics of technology and civilization. For example, Clas Bergman argued that it was a good thing that 'more people have been able to acquire a car', because it gave them 'increased opportunities to get out into nature'. The problem was that many car owners behaved dishonourably by unscrewing the licence plates and dumping old cars in the countryside. There they could stand 'in all their hideousness almost forever'. Bergman described this fly-tipping problem as destroying nature, which in this context meant that nature's aesthetic qualities were being nullified. There was no belief that the car wrecks constituted any direct danger to plant and animal life, much less to humans.[14]

However, in an article that appeared a year later, in 1961, the same Clas Bergman asserted that humans' economic encroachments into nature risked causing greater and more profound damage. He highlighted 'the ever-increasing use of toxins in the forestry and agricultural industries'. The problems were still insufficiently investigated, he said, but alarming US studies indicated that bird populations and breeding successes had fallen sharply in the wake of increased toxin use. Bergman emphasized that nature enthusiasts must become more vigilant. The use of toxins involved large sums of money, and its socioeconomic significance was rapidly increasing.[15] Another growing problem was oil spills. *Fältbiologen* reported that the newspapers and television had recently been full of reports about the 'terrible effects of oil damage on certain bird species'.[16]

During the early 1960s, Nature and Youth Sweden's interest in and commitment to nature conservation were significant. However, the association's activities were primarily aimed at helping its members

---

12 Per-Erik Tonell, 'Naturen samhället och vi', *FB*, 4 (1962), pp. 2–3; Roger Gyllin, 'Några synpunkter på naturvård – två inlägg', *FB*, 3 (1963), pp. 22–24.
13 Anne von Hofsten, 'Annes syn på saken', *FB*, 2 (1961), p. 2.
14 Clas Bergman, 'Höstfunderingar', *FB*, 3 (1960), p. 2.
15 Clas Bergman, 'Varför djurskydd?', *FB*, 3 (1961), p. 4.
16 Bredo von Bornstedt, 'Om oljeskadorna', *FB*, 2 (1962), p. 6.

to experience and learn more about nature. For that reason, these activities chiefly involved hiking, camping, and excursions. *Fältbiologen* published recurring reports from these events. It was with one such report, from 1964, that one member of the association, Erik Isakson, made his debut as a writer. The preamble described him as 'one of the Stockholm club's great fighters' and announced that he was going to 'lead the Stockholm district's toughest camp' during the summer. This was a tent camp in northernmost Sweden, with planned challenging hiking trips to Finland, Norway, and the Russian border. In a letter to the editors Isakson said that, as early as the beginning of May, he had gone north to 'leave the noise behind' and study the arrival of birds in mountains and wetlands. The letter described the species of birds he had seen and heard in detail.[17]

The next issue included a long report from the summer camp. Isakson described how twelve young members of Nature and Youth Sweden had travelled from Stockholm on 10 June and not returned home until 11 July. All the campers were boys in their late teens. A photograph in the magazine showed them posing for the camera carrying heavy backpacks. Isakson stressed the many difficulties they had overcome together: hunger, cold, and severe weather. One of the participants had fallen on sharp rocks and spat out 'teeth and blood'. On another occasion they had encountered Finnish soldiers with machine guns, who had taken them 'the shortest way back to civilization'. The adventure-filled stories were combined with detailed descriptions of nature and birdlife. The article was a playful display of interest and knowledge presented in a light, humorous way. Isakson said that the young adventurers always slept for a long time, because otherwise they would not have 'had the strength to see all these birds'.[18]

Erik Isakson was hardly a typical member of Nature and Youth Sweden. The vast majority of its members did not go to month-long summer camps in the wilderness. However, Isakson appeared regularly in *Fältbiologen*'s columns from 1964 onwards. Belonging to the inner circle of the Stockholm club, he had close contacts with the editorial staff. Consequently, he became visible to members in the rest of Sweden as well. Through his descriptions of nature and travel, he manifested many of the association's highest ideals:

---

17 Erik Isakson, 'Lappländsk vårvinter', *FB*, 3 (1964), pp. 13–15.
18 Erik Isakson, 'Nordkallet runt', *FB*, 4 (1964), pp. 10–15.

a desire for adventure, an interest in nature and the wilderness, activity and commitment, plus deep special knowledge about animal and birdlife.

As the 1960s progressed, the association combined its interest in experiencing and studying nature with increasingly diligent nature-conservation efforts.[19] In the wake of Rachel Carson's *Silent Spring* and the Swedish biocide debate, the association launched a campaign called 'Samla kalla fakta' [Collect Hard Facts]. It amounted to sending dead animals to the National Veterinary Institute for autopsy. The members' aim was to help researchers obtain scientific evidence for the harmful effects of biocides. Individual members were encouraged to provide real assistance 'by submitting *all* the dead things you find in nature'.[20] The campaign ran for two years (1963–1965) and marked the beginning of a phase of more direct commitment to nature-conservation issues.[21]

On Sunday 27 March 1966, another significant direct action occurred: the 'Clean a beach' event on the west coast of Skåne in southernmost Sweden. It was initiated by the Skåne district's sub-management working committee and a number of local clubs. They had been inspired by the Swedish Society for Nature Conservation's national campaign against littering in nature, *Håll Sverige Rent* [Keep Sweden Clean], which had been run several times in the course of the 1960s. The Nature and Youth Sweden action was supported by local landowners who provided tractors, trailers, and trucks. The regional *Håll Skåne Rent* campaign supplied paper bags to hold the rubbish.

Prior to the action, members of Nature and Youth Sweden made an appeal via the newspapers and radio to encourage 'the otherwise so passive city dweller to come out and make an active contribution to a more beautiful countryside'. They also sent an appeal, signed by Skåne's Province Governor Gösta Netzén, to 'everyone who might be interested in helping'. After the event, *Fältbiologen* reported that about 300 people had braved rain, wind, and cold to clean up the almost 20-km-long coastline between Landskrona and Barsebäck north of Malmö. The beaches had been 'veritable rubbish dumps' filled with 'plastic in all forms, bottles, petrol drums, etc'. The result

---

19 Håkan Agvald, 'Naturvårdare', *FB*, 4 (1965), pp. 22–23.
20 Olle, 'Gott nytt…', *FB*, 1 (1964), p. 2.
21 Hans-Georg Wallentinus, 'Den tysta våren redan på väg!', *FB*, 3 (1963), pp. 2–3; Hans-Georg Wallentinus, 'Samla kalla fakta', *FB*, 4 (1965), p. 24.

was described as fantastic. In five hours and 'in abysmal weather conditions, 2,300 sacks had been filled and the coast cleared of some 700 tonnes of debris'.

Nature and Youth Sweden's cleaning operation was described as the biggest ever in Skåne, and it received significant attention in the press and broadcast media. This media coverage was in turn extensively reported in *Fältbiologen*, which devoted a double spread to press clippings and photographs of the event. The newspaper headlines described a 'Huge effort by nature people' and an 'army' of Nature and Youth Sweden members. The photographs showed children and young people collecting plastic and dragging bags of rubbish, soaked to the skin.[22]

This cleaning action in Skåne in the spring of 1966 was in many ways characteristic of the association at this time. The young people were shown to be capable, responsible, active, and enterprising. Through their knowledge of, and feelings for, nature, the association's members wanted to help society. They did not oppose any kind of establishment; on the contrary, they were happy to cooperate with, and be variously supported by, the adult world. Their activities and opinions harmonized well with those of the Swedish Society for Nature Conservation. Furthermore, the action shows that nature conservation in the mid-1960s was largely about enhancing the beauty of nature. It should be preserved so that it could be experienced. There was no social criticism, no political agenda, no imminent threat to humanity. But all this was about to change.

Nascent environmental activism, 1967–1969

At the beginning of 1967, Thomas Söderqvist became the new editor of *Fältbiologen*. His ambition was to make the magazine 'a provocative agency for debate', and he wanted every member of Nature and Youth Sweden to be a potential 'habitat warden'. Other members of the association's national council expressed similar views. There was talk of 'raising the standard of debate', providing 'ecological education', and turning members into 'nature-conservation propagandists'.[23] These growing ambitions to recast the association's activities on the part of the national council coincided with the increasing momentum of the Swedish environmental debate.

---

22 Anders Rünow, 'Att städa en strand', *FB*, 2 (1966), pp. 7–11.
23 Anon., 'Sällskapet för inbördes beundran', *FB*, 1 (1967), p. 14.

The definitive breakthrough occurred in the autumn of 1967. Scientific warning voices, for instance those of Hans Palmstierna and Svante Odén, were making a great impact. *Fältbiologen*'s editorial staff intervened as well. One Saturday night, in early November, they went out into the Sergels Torg square in central Stockholm with tape recorders to ask young people what they felt and thought about nature conservation. The answers they received mainly focused on litter in nature. This angle was opposed by the editors. For them, nature conservation was about so much more than 'single-use glass and milk packaging'. It included the expansion of hydroelectric power on the Vindel River, biocides in agriculture, consumer responsibility, and the conflicts between environmental protection and economic growth. Consequently, the *Fältbiologen* editors were now critical of 'the SNF propaganda' against littering, claiming that it had given the public an overly narrow understanding of what environmental and nature conservation was all about.[24]

Outspoken criticism of the Swedish Society for Nature Conservation and *Håll Sverige Rent* was something new in *Fältbiologen*'s columns. It marks the beginning of a process of growing politicization and a more independent involvement in the environmental debate. In the summer of 1968, Wolter Arnberg criticized the incumbent Social Democratic government. He asked what difference it made that the government had adopted an ambitious environmental policy programme when it continued to make environmentally destructive decisions in practice. Arnberg stressed that the government, in direct conflict with nature-conservation interests, had given the go-ahead for a new airport in the Sturup area in Skåne and for industrial exploitation of the Väröbacka area in Halland. 'The young generation', he wrote, 'has the right to demand not to have to bear the burden of a society that has to be given artificial respiration via a contaminated environment.' All exploitation must be conducted 'with a view to the future'. A 'hunger for prestige and short-term profit interests' should not be allowed to 'crush the very foundations of our generation'.[25]

In the same issue, Mats Segnestam drew a gloomy picture of Sweden's 'many and big' nature-conservation problems. He cited the use of biocides in agriculture, the sulphur dioxide in the air, the

---

24 Thomas Söderqvist, 'Miljövård på Sergels Torg en lördagskväll', *FB*, 5–6 (1967), pp. 22–24.
25 Wolter Arnberg, 'Vindelälven åter hotad', *FB*, 3 (1968), p. 2.

mercury-poisoned lakes, gravel pits becoming depleted, and the difficulties in finding appropriate sites for 'nature-destroying facilities'. The situation demanded 'planning and *long-term thinking*' (italics in the original). However, he saw no sign of these. His list of conservation problems was only a small sample, he claimed, adding, '[a] complete list would be endless!'.

The extensive catalogue of problems was followed by a review of legislation, government bodies, nature-conservation organizations, and mass-media reporting. In these areas there were some hopeful signs. Segnestam emphasized that 'the average Swede' had probably 'begun to realize that nature conservation problems exist and perhaps suspect that they are serious'. The politicians had also realized 'that the general public has begun to wake up', and the political parties were therefore trying to 'surpass one another with ambitious environmental programmes'. If this could 'be put into practice, that would of course be excellent'. Segnestam hoped that the next generation would also have 'greater opportunities to absorb nature-conservation thinking' from an early age. The school system's curricula were being updated, and over the past year 'a tidal wave of podium debates, lectures, conferences, courses, and study circles on nature conservation has swept over us'. The situation was serious, but enlightened public opinion was gaining strength. Segnestam concluded that 'Our nation [literally 'Mother Svea', an expression that serves as a national symbol in Sweden] must not and cannot turn a deaf ear any longer!'[26]

The articles by Arnberg and Segnestam were typical of the new socially critical position that Nature and Youth Sweden began to adopt in the late 1960s. The association's leading representatives increasingly turned against politicians and other people in power, including the National Environment Protection Board. Nature and Youth Sweden did welcome the founding of the Board, but quickly became a critical monitor of its activities. The existence of a national environmental policy establishment in Sweden as early as 1967 was very important to Nature and Youth Sweden and the emerging environmental movements in the country: it gave them a clear and legitimate opponent. This position enabled them to share in, and agree with, much of the general criticism of the establishment that

---

26 Mats Segnestam, 'Se – hur skönt landet ligger', *FB*, 3 (1968), pp. 13–15. For studies of national planning in the 1960s, see Katarina Nordström, *Trängsel i välfärdsstaten: Expertis, politik och rumslig planering i 1960- och 1970-talets Sverige* (Uppsala: Studia Historica Upsaliensia, 2018).

was gathering force during the late 1960s. As a result, Nature and Youth Sweden became part of the new political culture of the period.[27]

However, it took some time for Swedish society as a whole to discover that Nature and Youth Sweden had changed. As mentioned earlier, the Sergels Torg demonstration in March 1969 aroused surprise. But for Wolter Arnberg at the speaker's podium, criticism of society and the establishment was nothing new: he and others had been expressing these views in *Fältbiologen*'s columns since 1967. The Sergels Torg protests were aimed at all forms of 'shortsighted planning' by Swedish authorities. In particular, Arnberg attacked 'Valfrid Paulsson and his National Environment Protection Board for their sickeningly watered-down nature-conservation policy'. He especially criticized the Board's putting the interests of the tourism industry before those of nature protection. 'Does the National Environment Protection Board want to build cabins in the national parks?' he demanded.[28]

Nature and Youth Sweden's demonstration on 16 March 1969 attracted attention not only from the press and broadcast media but also from the state-owned energy company Vattenfall, which was responsible for expanding hydroelectric power on the northern rivers. In conjunction with the Sergels Torg demonstration, the company's chief press and information officer made contact and asked representatives of Nature and Youth Sweden to come to Vattenfall's office the next day to discuss the issues. The proposal was accepted, and on the evening of 17 March there was a panel debate led by Eskil Block, a scholar specializing in future studies. Nature and Youth Sweden's representatives Wolter Arnberg and Lars-Erik Liljelund were pitted against Vattenfall's director-general Erik Grafström and civil engineer Väinö Wanhainen. The four-hour debate was recorded by Radio Sweden, which broadcast a feature about it on the late TV news.[29]

A week later, members of Nature and Youth Sweden, led by its chair, Wolter Arnberg, visited Minister for Agriculture Ingemund Bengtsson. They handed him a letter and expressed their views. In

---

27 Salomon, *Rebeller i takt med tiden*; Östberg, *1968 – när allting var i rörelse*; Östberg and Andersson, *Sveriges historia*; Jørgensen, *Transformation and Crises*; Berggren, 68.
28 Wolter Arnberg, 'Tal framfört på Sergels Torg den 16.3 1969 vid Fältbiologernas demonstration mot vattenkraftutbyggnaden i fjällen', *FB*, 3 (1969), p. 5.
29 Erik Isakson, 'Sluta reglera våra älvar', *FB*, 3 (1969), pp. 5–6.

Fältbiologen Erik Isakson reported that the minister 'listened kindly, but had some trouble understanding our pessimistic view of the future and of environmental protection'. Bengtsson wanted the young people to be optimistic about the future. He emphasized that much had already been done in Sweden and pointed out that no other country in the world had introduced such stringent restrictions on environmental destruction. Isakson was not convinced. He was certain that the future would prove the leading Social Democratic politicians wrong. The situation, he said, was far more serious than they imagined.[30]

Nature and Youth Sweden's increasingly fierce criticism of the establishment had one exception: the scientific research community, with which the association had always had close ties. In the late 1960s several of its leading members, including Thomas Söderqvist and Wolter Arnberg, had been university students with research ambitions. The association was especially favourably disposed towards the new environmental debaters. Both Hans Palmstierna and Nils-Erik Landell were given space to write for *Fältbiologen*. The fact that Palmstierna was an active Social Democrat was not a disadvantage. He was a scientific voice whom the nascent environmental activists wholeheartedly supported.[31]

Even so, this growing engagement with environmental policy did not mean that field studies and outdoor life were given less priority. These traditional activities continued to be central, and *Fältbiologen* regularly published reports and informative texts along those lines. In 1968, for example, Erik Isakson wrote an in-depth guide to the equipment that a Nature and Youth Sweden member would need in order to survive in Sweden's mountain areas. It was vital to know about everything from tents and backpacks to clothes and provisions.[32] Later he shared his experiences of various kayaks and types of ski equipment.[33]

In 1969, Erik Isakson became editor of *Fältbiologen*. In his first editorial, he described his 'happy youth' when he had spotted 350 species of birds. Now, though, he had 'discovered that there are other things than bird species in nature', and that 'Stockholm is

---

30 Ibid., p. 6.
31 Hans Palmstierna, 'Undvik vägen utan återvändo', *FB*, 4 (1968), pp. 3–5; Nils-Erik Landell, 'Hur ska vi ändra framtiden', *FB*, 6 (1969), pp. 4–5.
32 Erik Isakson, 'Skogsliv till fjälls', *FB*, 2 (1968), pp. 3–6.
33 Erik Isakson, 'Kanoter', *FB*, 5 (1969), pp. 11–13; Erik Isakson, 'Skidor och bindningar', *FB*, 6 (1969), p. 16.

dirty and destructive to humans'. He criticized the 'human spirit of development', stressing that all scientific and technological advances brought 'unsuspected disadvantages'. Today's situation could perhaps be mastered 'if we did more than today's petty efforts'. At the same time, though, 'new so-called advances were constantly taking place, which in their turn had unsuspected effects'. The new editor advised his readers to go and visit the mountains before they were destroyed.[34]

Isakson's criticism of civilization ran like a central thread through *Fältbiologen* during 1969, and his own texts came to revolve more and more around 'Greenland'. For him, it became a symbol of 'the free life far from the devastating effects of modern technology'. Nature and Youth Sweden's debates with Vattenfall and Valfrid Paulsson in the spring of 1969 made him 'rather sad'. In his view, 'the ruling politicians and the sausage-stuffed experts' were unyielding in their belief 'in the redeeming effect of technology for humanity'. Continued debate was pointless and had about the same effect as stamp collecting. Isakson hence urged his readers to strive to be independent of technology and to turn their backs on modern society.[35]

Erik Isakson was not alone in raising his voice in 1969. Under his editorship, critical debate and social criticism were given more and more space in *Fältbiologen*. Writers began to depict historical developments in sombre colours. Kristoffer Andersson described how industries had 'popped out of the earth like poisonous mushrooms' and how environmental destruction was now coiling its way 'like a hungry Midgard serpent around the entire globe'.[36] Ulf Rooth argued that Nature and Youth Sweden had to keep up with the times and update its goals. Field studies and peer education were no longer sufficient. The association's task should be 'to spread information about environmental destruction to the people, and to present their and our views to the politicians'.[37]

Lifestyle issues also began to be discussed at this time. Marianne Reini maintained that it was important for members of Nature and Youth Sweden to practise what they preached. 'How can we criticize

---

34 Erik Isakson, 'Gifterna åt människan', *FB*, 1 (1969), p. 2.
35 Erik Isakson, 'Variation nummer fyra', *FB*, 3 (1969), pp. 2–3. 'Sausage-stuffed' is a literal translation of a Swedish expression, signifying massive soulless cramming.
36 Kristoffer Andersson, '?', *FB*, 2 (1969), p. 2.
37 Ulf Rooth, 'Naturen, Fältbiologerna och Samhället', *FB*, 6 (1969), pp. 2–3.

Emergence of the modern environmental movements

environmental destruction as long as we ourselves use the products that are being manufactured?' she demanded.[38]

The magazine's former editor, Thomas Söderqvist, said that he had moved out to a smallholding in the country – an act he wanted to 'propagandize for'.[39] As usual, Erik Isakson's was the most radical voice. He talked about moving to Greenland and also of 'mustering a guerrilla army and resorting to violence'.[40] Much had changed since the good-natured cleaning action in Skåne in the spring of 1966.

Radical social critics, 1970–1972

Nature and Youth Sweden began the 1970s by demonstrating on the streets of Norrköping (a city some 160 km south-west of Stockholm, with a population of about 140,000) and adopting the association's first political programme. That happened at the annual meeting in early January when more than a hundred members gathered to socialize, go on excursions, make placards, and attain shared 'doctrines and values'. The new programme announced that we humans were 'obliged to preserve our limited natural environment for the sake of future generations' and that the population explosion had 'developed into a catastrophe'. It also asserted that the Swedish authorities either did not realize the seriousness of the situation or failed to accept the 'consequences of their realization'. The measures that the politicians had implemented and were planning were dismissed as 'completely insufficient'.[41]

The overall goal of Nature and Youth Sweden's political programme was that 'our civilization' should not 'make a rich and varied biological life on Earth impossible in the future'. To ensure this would not happen, the association would 'spread knowledge and awareness of nature-conservation issues', provoke debate, demonstrate, write letters, and carry out direct actions. Members should 'assert their views according to their conviction', whether

---

38 Marianne Reini, 'Inlägg i debatten om Fältbiologerna', FB, 6 (1969), p. 3.
39 Thomas Söderqvist, 'Tankar från torpet', FB, 4 (1969), p. 23.
40 Erik Isakson, '?', FB, 2 (1969), p. 2; Erik Isakson, 'Detta nummer', FB, 6 (1969), p. 2.
41 Anon., 'Program för Fältbiologernas naturvårdsverksamhet antaget av årsmötet 1970', FB, 1 (1970), p. 3; Rolf Jacobsson, 'Årsmötet 1970', FB, 1 (1970), pp. 20–21.

or not they agreed with the views of any political party or gave positive or negative publicity to the association.[42]

The 1970 annual meeting developed into a manifestation of the transformation that had taken place in Nature and Youth Sweden. Rolf Jacobsson proudly described in *Fältbiologen* how 'flaming torches lit up our placards', which 'shouted out terrifying and accusatory truths to the sceptical world around us'. It was reported that the local press had portrayed the demonstration as 'a powerful indictment of society'.[43] In the same issue, Erik Isakson pointed out that 'the material standard of living must be lowered' if 'life on Earth is to continue'.[44]

The growing political involvement within Nature and Youth Sweden had a bias towards the political left. The clearest example of this was Thomas Söderqvist. In the autumn of 1970, he encouraged all members of the association to contact left-wing political action groups. For him, the environmental crisis was interwoven with a profound social crisis of global proportions. It was therefore pointless to look for technological solutions to the environmental problems. A radical transformation of society was required. 'The real enemy', he argued, 'is the inhumanity that emerges in the industrialized affluent societies.' This enemy could be labelled 'capitalism, bureaucracy, imperialism, or whatever you want'. The basic problem was the same: the blind pursuit of material growth. This could only be attacked politically. For that reason, he believed that members of Nature and Youth Sweden should ally with the New Left.[45]

Söderqvist's contribution to the debate did not go unchallenged. Outraged members of the association contacted the editorial department and questioned 'the validity of conducting political propaganda in the magazine'.[46] As editor, Erik Isakson stressed that Söderqvist's post was not written 'in the association's name'.[47] At the same time he afforded space to a text by Wolter Arnberg, chair of the association, about its new political programme. Arnberg summarized and commented on the 'doctrines and values' adopted at the annual meeting.

---

42 Anon., 'Program för Fältbiologernas naturvårdsverksamhet antaget av årsmötet 1970', p. 3.
43 Jacobsson, 'Årsmötet 1970', p. 21.
44 Erik Isakson, 'Förbjud snöskotrarna', *FB*, 1 (1970), p. 3.
45 Thomas Söderqvist, 'Politisera miljövårdsdebatten!', *FB*, 3–4 (1970), pp. 30–31.
46 Anders Fridell, 'Politisk propaganda', *FB*, 5 (1970), p. 29.
47 Erik Isakson, 'Från när och fjärran', *FB*, 5 (1970), p. 25.

He stressed that 'environmental protection is politics' and that the association's task was 'to try to increase your knowledge about your surroundings and to create a pressure group for the care of our environment'. Nature and Youth Sweden would be an active voice in the environmental debate and one that expressed political views based on 'biological facts'.[48]

Nevertheless, opinions were divided as to how the association should proceed. Torbjörn Kronestedt felt that Söderqvist was advocating an overly idealistic position. To bring about real change, people needed to be realistic and willing to cooperate both with politicians and with the scientific research community. 'A general mood of dissatisfaction with a global set of problems produces little result', he declared.[49] Söderqvist replied that Kronestedt was being deceived. 'The powerful authorities and the powerful business community have long tried to convince us that society can only function in this way.' The school system and mass media had been indoctrinating us all 'since we were toddlers'. The Swedish welfare state was built on 'the doctrine of growth'. This approach was no longer defensible because 'increased, uncontrolled economic growth poses the greatest threat to positive ecological interaction'. Capitalism was the problem according to Söderqvist, who added that only the left-wing groups shared this insight and offered alternatives.[50]

As we have seen, *Fältbiologen* became a forum for internal debate about political ideology in the early 1970s. Conceptually, there was also a shift in emphasis from 'nature conservation' to 'environmental protection'. The changes coincided with the association's new valorization of knowledge about how society functioned. Environmental protection was considered to be fundamentally a social and political issue. It was therefore no longer sufficient for members of Nature and Youth Sweden to possess in-depth knowledge about bird species and outdoor life. The magazine's editorials recommended a number of books, published by the state-owned Allmänna förlaget [General Publishing], which could be read in study groups. The books reflected 'society's view of the environmental-protection problem and how it should be solved', which was important for members to know before they contacted various authorities.[51]

---

48 Wolter Arnberg, 'Fältbiologerna och deras politik', *FB*, 5 (1970), pp. 25–26.
49 Torbjörn Kronestedt, 'Först vettig "miljövårds"-debatt här hemma', *FB*, 5 (1970), pp. 26–27.
50 Thomas Söderqvist, 'Död åt vinstintressena', *FB*, 5 (1970), p. 28.
51 Lars-Erik Liljelund, 'Vad gör samhället för vår miljö', *FB*, 4 (1971), p. 78.

The association's social criticism was primarily aimed at the Social Democratic establishment and in particular at the director-general of the National Environment Protection Board, Valfrid Paulsson, and the minister for agriculture, Ingemund Bengtsson. Nature and Youth Sweden regarded the two men as opportunistic defenders of modern industrial society. A satirical series published in 1971 described Paulsson as 'the little watchman of all nature'. He saw all the environmental destruction going on and reacted with lightning speed. But not by stopping the transgressions and punishing the offenders. Instead, he made sure to track down all the 'illegal destroyers of the environment' and give them a permit.[52]

Lars-Erik Liljelund argued that the National Environment Protection Board was not very 'non-profit-focused'. He said that the Board kept making decisions that went against environmental-protection interests. Liljelund found it remarkable that Paulsson 'almost never publicly criticizes the environmentally hazardous activities of various companies and municipalities'. What he did do, according to Liljelund, was to repeatedly make 'rancorous jabs at the large group of people who feel powerless and concerned about the future and the environment we will have'. Who were Paulsson's true opponents, Liljelund asked – the actors who were destroying the environment or the ones that were trying to protect it?[53]

Nature and Youth Sweden also expressed its political commitment in the form of direct actions. These were designed to provoke media commotion about various environmental protection problems, with the aim of influencing public opinion and promoting change. In the wake of successful actions, *Fältbiologen* would publish reports and press clippings. The stated purpose was to inspire imitation and to spread knowledge within the association about how environmental-protection actions could best be carried out. It is noteworthy that the association was employing these deliberate media strategies at such an early stage. Over time, these methods became characteristic of the international environmental movement, not least through Greenpeace, which was formed in Canada in 1971.[54]

---

52 Lars Olsson, 'VIP=Very Important Paulsson', *FB*, 1 (1971), p. 30. Cf. Per Jadéus, 'Oberättigad kritik?', *FB*, 2 (1972), p. 47 (a humorous sketch of Valfrid Paulsson).
53 Lars-Erik Liljelund, 'Hej', *FB*, 5–6 (1971), pp. 82–83.
54 Zelko, *Make it a Green Peace!*

In April that same year, the Nature and Youth Sweden club in the small town of Mariestad carried out a direct action over rubbish, with the intention of raising awareness of local waste-management problems. The club's members acted as investigative journalists, and Bo Landin proudly reported how they had tricked those responsible into exposing bad practices. Among other things, the members had gone to the local landfill site and pretended to be 'pupils interested in photography who perceived something beautiful in the rubbish'. On another occasion, they had pretended to be one 'Mr Gustafsson' who wanted to scrap a car. They had then found out that the private company operating the landfill had 'a contract with an old fellow in the forest'. So here was one reason why there were so many scrap cars in the countryside!

The club's revelation of the abuses was carefully orchestrated. Its members secretly prepared an exhibition at the local bank branch, wrote a letter to the local health authority, and ensured that the local press was on standby. A relevant circumstance here is that two of the club members worked for the local newspaper, *Tidning för Skaraborgs län*. This allowed them to steer the press coverage and the subsequent debate. Landin reported that the revelation of the mismanagement of the rubbish tip had led to 'chaos and consternation' in Mariestad. This was followed by new headlines about the scandal, complaints being filed to the police, and – soon after – actual changes in how the waste was managed. Nature and Youth Sweden itself considered the end result of the action to be 'sensational'.[55]

In the spring of 1972, Nature and Youth Sweden joined forces with Miljögruppernas riksförbund (MIGRI) [the National Association of Environmental Groups] and Jordens Vänner [a co-founder of the Swedish branch of the international organization Friends of the Earth] to carry out a nationwide campaign against single-use packaging. The week-long endeavour included distributing flyers, holding demonstrations, and collecting signatures. The so-called 'packaging action' attracted the most attention. This involved members of Nature and Youth Sweden going out into the countryside throughout Sweden and collecting discarded single-use packages. These were then wrapped up and posted to the Swedish government, the Riksdag, the National Environment Protection Board, and the leading Swedish packaging manufacturer PLM. *Dagens Nyheter* reported that 160,000 beer cans had been sent by post. Valfrid Paulsson condemned the action

---

55 Bo Landin, 'Sopaktion i Mariestad', *FB*, 5–6 (1971), pp. 109–113.

as 'completely insane', adding that 'we environmentalists must beware of appearing as foolish romantics'.[56]

Of course, Nature and Youth Sweden took a different view of the matter, regarding the campaign as a 'complete triumph'. About a hundred of the association's clubs, plus about fifty other environmental-conservation groups, were reported to have taken part in it. In addition, 70,000 signatures were reportedly collected against 'the single-use hysteria'. Nature and Youth Sweden felt that the association had imparted 'full speed to the debate over the waste of raw materials'. It added that PLM had been forced to admit that it was driven solely by profit motives and did not care about the environmental consequences of its activities.[57]

Nature and Youth Sweden's radical criticism of the establishment was also directed at the upcoming UN conference in Stockholm. Lars-Erik Liljelund joked that the environmental-protection conference was a 'weapon of total annihilation against the global environmental destruction'. He objected to the exclusion of the general public, not least 'rebellious elements' in it. The conference only existed so that 'the real decision-makers' could meet and discuss 'the problems they themselves have caused'.[58] Christina Skarpe argued that no improvements came from adopting declarations and resolutions. She felt that the main value of the conference was 'the publicity it will give to environmental protection'. She also regretted that the conference focused on 'finding ways to combat the symptoms of environmental problems, rather than preventing their emergence'.[59]

Nature and Youth Sweden was indirectly represented at the Stockholm Conference by the International Youth Federation for the Study and Conservation of Nature (IYF, founded in 1956). In the summer of 1971, the above-mentioned Christina Skarpe and Bo Landin had participated in the organization's conference in Canada. Their experience from that event was that international environmental protection work was complex and difficult. Cooperation between participants from the industrialized and developing countries was particularly tricky. In his report, Landin wrote that it boded ill for the future that the latter's representatives were not interested 'in hearing how we are now trying to solve problems which they

---

56 Christina Kellberg, 'Paulsson rasar över ölburkarna: "Ansvarslöst av fältbiologerna"', *DN*, 18 April 1972.
57 Bo Landin, 'Fältbiologerna till attack!', *FB*, 3 (1972), pp. 62–64.
58 Lars-Erik Liljelund, 'Hej', *FB*, 4 (1971), p. 4.
59 Christina Skarpe, 'Vad händer i Stockholm?', *FB*, 2 (1972), pp. 35–36.

don't have yet and learning from us'. Instead, the developing countries' representatives were more eager to assert 'their position in the world and their independence'.[60] The lecturing attitude expressed in Landin's report is likely to have been one of the reasons for the difficulties in cooperation.

Even so, members of Nature and Youth Sweden who wanted to be directly involved in the Stockholm Conference and in arrangements connected with it did have some opportunities. Christina Skarpe explained that two open meeting-places would be set up: 'Environment Forum' and 'Folkets Forum' [The People's Forum]. The former was organized by the UN Association of Sweden and the National Council of Swedish Youth (SUL), while the latter brought together newly founded environmental groups such as Jordens Vänner [Friends of the Earth], Dai Dong, and Powwow.[61] Nature and Youth Sweden did not assume any official position regarding the two forums, but for the association's radical elements the choice was not difficult. Folkets Forum offered a meeting place for environmental and establishment criticism with ties to the New Left.[62]

Nature and Youth Sweden and the environmental movements in Sweden

The transformation of Nature and Youth Sweden is an important part of the history of the Swedish environmental movements. What makes the association especially interesting is that it spans and connects several different lines of development. Nature and Youth Sweden was founded within an older nature-conservation tradition and was, as we have seen, firmly anchored in that tradition until the autumn of 1967. After that, a gradual politicization occurred, especially at the leadership level in Stockholm. In the early 1970s, more and more members of the association became environmental activists. The members' young age made knowledge about the global environmental crisis a particularly urgent matter for them. Environmental destruction and unfettered economic growth posed a direct threat: what was at stake was their own future.

---

60 Bo Landin, 'IYF', *FB*, 4 (1971), pp. 71–74.
61 Christina Skarpe, '... och var kommer vi in i bilden?', *FB*, 2 (1972), pp. 37–38.
62 Lars Gogman, 'Rödgrönt samarbete med förhinder', *Arbetarhistoria* 142.2 (2012), p. 48.

Even after this radicalization, however, older ideals rooted in nature-orientated Romanticism were very much alive within the association. This is how its magazine's editor, Roger Olsson, described what was behind his own involvement in the environmental debate in the summer of 1972: 'For me, it is the feeling for nature', he wrote, 'a feeling created by all the experiences of happiness nature has given me.' He hoped and believed that most members of Nature and Youth Sweden had the same motivation for their actions. 'Because it is a good foundation. One knows and feels that one is fighting for something incredibly valuable, something that is far more necessary than everything else.' In his view, the association's most important task was hence always to arouse feelings for nature in its members. If they also had time to hand out leaflets, scold Vattenfall, and send beer cans to the minister for agriculture, that was a good thing. But it was not as important as actually experiencing and studying nature. A field biologist's boots should be kept muddy.[63]

Roger Olsson's comments reveal the tension between an apolitical interest in nature and radical environmental activism. At the beginning of the 1970s, Nature and Youth Sweden encompassed both. The relative importance of different ideals and practices was not obvious. In this respect, too, there are interesting differences between Nature and Youth Sweden and its parent organization, the Swedish Society for Nature Conservation (SNF). The latter also underwent a revolutionary change in the years around 1970, as voices critical of civilization and economic growth became ever stronger.[64] Nevertheless, SNF continued to have close ties with established institutions in Swedish society. The adult members' commitment to environmental protection and nature conservation was not manifested in demonstrations and direct actions, and the political culture of the New Left did not have the same appeal to them.

Other environmental organizations did have a more obvious extra-parliamentary left-wing profile. They notably included the environmental groups and residents' associations in Sweden's major cities, who opposed concrete transformations of the urban environment. Among these, Göteborgs miljögrupp [Gothenburg's Environmental Group] and Alternativ stad [Alternative City] in Stockholm are especially well known. The latter was strongly involved in the so-called 'Battle of the Elms' in the spring of 1971, when protesters sought to prevent the felling of thirteen elm trees in the

---

63 Roger Olsson, 'Håll stövlarna leriga!', *FB*, 4 (1972), p. 74.
64 Anshelm, *Det vilda, det vackra och det ekologiskt hållbara*, pp. 60–72.

Kungsträdgården park in central Stockholm. (They succeeded; the elms are still standing.) The action received much media attention, and, like Nature and Youth Sweden's actions, it expressed sharp criticism against the establishment.[65] The residents'-association movement was loosely organized. It played a part both in the design of the physical urban environment and in the creation of a sense of community among people in a neighbourhood. The focus tended to be on local issues, such as opposing demolition or new construction.[66] For these organizations, environmental toxins, industrial exploitation of the wilderness, and global survival problems were not such key issues as they were for Nature and Youth Sweden or for the Swedish Society for Nature Conservation.

Environmental commitment in Sweden was also channelled through the established political parties, especially their youth wings. Most important in this context was the youth wing of the agrarian non-socialist centre party, Youth League of the Centre (CUF) [now Centre Party Youth]. CUF was involved in urban transformation, rural depopulation, and issues to do with global resources and justice. Its political vision was a decentralized society in which public authorities and technocratic experts had less say. Democracy should be built around strong local communities, and CUF therefore advocated an active regional and localization policy. Like Nature and Youth Sweden, CUF took the global environmental crisis very seriously and warned against blind trust in economic growth and technological progress. For obvious reasons, however, there was not the same radical critique of the capitalist system within CUF. It also distanced itself to some extent from the New Left's culture of political action.[67]

In the early 1970s, a number of new environmental organizations were formed in Sweden. They included Jordens Vänner [Friends of the Earth] and the national association of environmental groups,

---

65 Daniel Halldén, *Demokratin utmanas: Almstriden och det politiska etablissemanget* (Stockholm: Department of Political Science, University of Stockholm, 2005).

66 Ulf Stahre, *Den alternativa staden: Stockholms stadsomvandling och byalagsrörelsen* (Stockholm: Stockholmia, 1999). For studies of the Swedish alternative movement see Kristoffer Ekberg, *Mellan flykt och förändring: Utopiskt platsskapande i 1970-talets alternativa miljö* (Lund: Department of History, University of Lund, 2016).

67 Carl Holmberg, *Längtan till landet: Civilisationskritik och framtidsvisioner i 1970-talets regionalpolitiska debatt* (Gothenburg: Department of History, University of Gothenburg, 1998).

Miljögruppernas riksförbund (MIGRI).[68] The latter was led by Björn Gillberg, a young researcher into heredity who was beginning to make a name for himself in the autumn of 1969 with the debate paperback *Hotade släktled* [Threatened Generational Descent]. Gillberg especially focused on environmental toxins. He argued that too much attention was being given to direct damage by toxins whereas their long-term effects on genetic material, and thus on future generations, were being ignored.[69] The following year he went on the offensive against the detergent industry. Gillberg warned that laundry detergents contained potentially carcinogenic substances and that the authorities were defending the interests of big business. He made the attack in and through *Dagens Nyheter*.[70]

Gillberg experienced his big media breakthrough on 12 October 1971 with the prime-time broadcasting by Swedish TV2's environmental department of the programme 'Han kan bara inte hålla käften' [He just can't shut up]. It portrayed Gillberg as a young, angry environmental debater who was not afraid to speak truth to power. Particularly striking was the fact that he washed his shirts with a coffee whitener called Prädd. The product quickly vanished from the market and Gillberg became, in the words of the historian of science and technology Per Lundin, 'a media darling'.

Lundin has analysed Gillberg's actions in the environmental debate of the early 1970s and their consequences. He points out that at the time of his breakthrough, Gillberg was employed as a project researcher at the Swedish University of Agricultural Sciences (SLU) outside Uppsala. He therefore possessed the scientific legitimacy necessary for him to be able to act as an expert in the media. Gillberg proved to be skilled at handling media logic. He made dramatic moves, issued blunt messages, and used instructive examples as well as simple language. His cocksure style led to attention and headlines but encountered a sceptical attitude in the research community.

---

68 Peter Larsson, 'Miljörörelsen', in Mats Friberg and Johan Galtung (eds), *Rörelserna* (Stockholm: Akademilitteratur, 1984), pp. 249–263; Magnus Boström, *Miljörörelsens mångfald* (Lund: Arkiv, 2001).
69 Björn O. Gillberg, *Hotade släktled: Genetisk bakgrund till några viktiga miljövårdsproblem* (Stockholm: Natur & kultur, 1969); Björn O. Gillberg, 'Hotet mot arvsmassan', *DN*, 20 November 1969.
70 Björn O. Gillberg, 'Naturvårdsverket spelar tvättmedelsföretagen i händerna', *DN*, 5 June 1970; Erik Brandt, 'Kemikontoret om NTA-stoppet: Nej, vi gick inte företagens ärenden', *DN*, 13 June 1970; Björn O. Gillberg, 'Visa byken om tvättmedel offentligt', *DN*, 3 July 1970; Barbro Josephson, 'Optiskt vitt – den nya faran?', *DN*, 24 August 1970.

Gillberg's media success hence undermined his scientific legitimacy, which made his media position untenable in the long term. In order to deal with the situation, he sought to strengthen and consolidate his position within MIGRI. The movement raised funds for his research and set up the Miljöcentrum [Environmental Centre] foundation. In 1972, the foundation began publishing its own periodical *Miljö o framtid* [Environment and future].[71]

At the time of the Stockholm Conference in 1972, Björn Gillberg's position in the media and in the Swedish environmental movement was still strong. Scientifically, however, he was marginalized. The big packaging action carried out by Nature and Youth Sweden and MIGRI against PLM was not something that was highly valued in the world of research. The many controversies surrounding Gillberg were typical of him as an environmental debater. Seen from a slightly longer time perspective, however, they were also typical of the Swedish environmental debate in the early 1970s. In this respect, an important dividing line runs between this debate and that of the late 1960s, especially of the breakthrough stage. In the autumn of 1967, the social circulation of knowledge in Sweden was characterized by relative consensus. At that time, it was appreciated that an actor like Hans Palmstierna built bridges between the worlds of science, media, and politics. Nor did the interest in environmental issues have any obvious political colouring at that point.

In the early 1970s, the situation was different. The political culture was transforming, and there were conflicts regarding the nature and status of factual knowledge. This chapter has shown how young environmental activists turned against the established society. But the cracks existed in the adult world as well. As environmental issues became politicized, the lines of conflict between various interest groups in society became more apparent. This was also true with regard to the sciences, and the ensuing chapter presents two open conflicts which attracted much attention in Sweden in the early 1970s.

---

71 Per Lundin, '"Han kan bara inte hålla käften": Björn Gillberg, lantbruksvetenskapernas medialisering och 1970-talets miljödebatt' (unpublished manuscript).

7
Conflicts and media storms, 1971–1972

On 1 September 1971, the international environmental trade fair 'Luften, larmet och vi' [The air, the alarm, and us] opened in Jönköping in the province of Småland. It drew crowds from far and near. The local press reported the presence of about eighty Japanese visitors and told readers that German and English 'were buzzing busily' among the participants.¹ Despite this foreign element, the inaugural speaker, Axel Iveroth, managing director of the Federation of Swedish Industries, gave his speech in Swedish. He explicitly addressed a national audience. 'It is no exaggeration', he said, 'to assert that our people are now animated by a colossal environmental ambition.' Perhaps it was even 'the biggest popular movement ever in the field of environmental protection'.²

Iveroth emphasized that commitment to the environment in Sweden encompassed both the external and the internal environment. The former involved the air, water, forests, and land; the latter concerned the working environment. This distinction was typical of the time, and it reflected an ongoing political struggle over the direction of the environmental debate. The Social Democrats argued that occupational-safety issues should be considered as important as nature-protection issues. This stance constituted an attempt to advance the battle lines within the environmental debate and also to gain support for the party's environmental policy from the trade-union movement. Unlike mercury-riddled pike, toxins on the factory floor were an issue behind which the labour movement could mobilize.³

---

1 Arthur Eriksson, 'Många utländska besökare', *Smålands folkblad* (*Sf*), 2 September 1971.
2 Axel Iveroth, 'Industrin och miljövården', p. 1, 452/2/14 (HP ARBARK).
3 Anshelm, *Socialdemokraterna och miljöfrågan*, pp. 32–36.

In his speech, Iveroth commented on this development and said that the external environment had long been at the forefront of the environmental debate. In recent months, however, occupational-safety matters had attracted increasing interest. In his view, the debate on that topic was characterized by intensity and emotions rather than by expertise. He singled out 'the latest outburst from one of our so-called environmental celebrities'. That celebrity was Hans Palmstierna. On behalf of the Swedish Factory Workers' Union, he had led an investigation into health risks in the Swedish chemical industry. His report had been presented at the Union's congress in Stockholm on 20 August 1971.

Speaking from the rostrum, Palmstierna had attacked both industry in general and the scientific community, accusing them all of not taking work-environment issues seriously. He also sharply criticized the fact that university-employed researchers conducted consultancy work for industry. An infected debate flared up in the press and Palmstierna was attacked from many directions, not least by researchers. In his own speech, Iveroth said that Palmstierna was no longer an alarm clock: he had lost the ringing tone he once had. 'He has nothing more to say, when, in order to get attention, he feels it necessary to use such crude expressions as those in his speech at the Factory Workers' Congress'.[4]

Iveroth's stand also caused much media commotion. On *Dagens Nyheter*'s front page, it was described as 'one of the most magnificent personal attacks in the Swedish debate in a very long time'.[5] Palmstierna chose not to reply directly. He told *Svenska Dagbladet* that he was sad and disappointed and felt that he had been misunderstood.[6] Writing privately, though, his tone was different. He described Iveroth's attack as 'politically daft'. He would let it stand on its own 'in all its shining, revealing glory'. Actually, Palmstierna maintained, Iveroth had done the labour movement a favour, because his speech had shown that the era of the class struggle was not over.[7]

The media storm around Palmstierna and Iveroth illustrates how fundamentally the dynamics of the Swedish environmental debate

---

4 Iveroth, 'Industrin och miljövården', pp. 1, 7.
5 Björn Berglund, '"Groteskt skvaller av hr Palmstierna..."', *DN*, 2 September 1971.
6 Anon., 'Palmstierna: Jag är besviken', *SvD*, 2 September 1971.
7 Letter from Hans Palmstierna to Yngve Persson, 5 September 1971, 452/3/6 (HP ARBARK).

had changed within the space of a few years. In the late summer of 1971, environmental issues were politicized – something they had not been during the breakthrough in the autumn of 1967. At that time, Palmstierna had referred to the need for information and enlightenment rather than speaking of a class struggle. But what were the consequences of this altered climate of debate? How was the circulation of knowledge and expertise affected? Was it still possible in the early 1970s for a scientist who sounded the alarm to become a unifying figure in the environmental debate? Or were the lines of conflict already drawn up and the positions locked?

In the late summer of 1971, this might have seemed to be the case. The conflict between Palmstierna and Iveroth ran very deep indeed. But the environmental debate at that time encompassed more actors, and also more topics, than the internal and external environments. The growing interest in the major global issues of the future was particularly significant. Where was humanity heading? Were humans becoming too many? How long would the natural resources last?

These types of question had been very present in the environmental debate in the autumn of 1967, not least thanks to Hans Palmstierna. In the autumn of 1971, though, other actors were tackling the survival issues. The person who came to have the greatest impact was Gösta Ehrensvärd, professor of biochemistry at Lund University and – like Palmstierna (whom he outranked, being a count whereas Palmstierna was a baron) – a member of the Swedish nobility. In the 1960s, he had gained a reputation as a popular-science writer; but his really big public breakthrough came in 1971 with the paperback *Före – Efter: En diagnos* [Before – after: a diagnosis]. Released in October, it quickly became a great commercial success, topping the bookstores' sales lists in the 1971 Christmas-shopping season.[8]

In his book, Ehrensvärd argued that the hyperindustrialized society of the 1970s would become a historical parenthesis. His calculations showed that the depletion of the Earth's limited resources, combined with accelerating population growth, would lead to a global crisis in about 2050. He predicted that there would then be centuries of famine and anarchy before a much smaller human race would return to living in an agrarian society at the level of eighteenth-century

---

8 This is based on David Larsson Heidenblad, 'Framtidskunskap i cirkulation: Gösta Ehrensvärds diagnos och den svenska framtidsdebatten, 1971–1972', *Historisk tidskrift* 135.4 (2015).

Sweden, with the addition of a few technological and chemical industries. By Western standards of the 1970s, the future standard of living would be low; but it would at least be stable in the long term. Industrial civilization's days were numbered – but not those of humanity.[9]

Ehrensvärd's prediction led to an intense debate about the future of humankind. The spark was ignited by nuclear physicist Tor Ragnar Gerholm's counterblast, *Futurum exaktum*, published in February 1972. However, the debate between and around Ehrensvärd and Gerholm was not coloured by any personal or party-political conflict. It focused on knowledge and its limits. What could scientists really say about the future? What validity did calculations, forecasts, and scenarios possess? These questions were particularly highly charged just before the imminent United Nations Conference on the Human Environment, which was to be held in Stockholm in June 1972.

The debates about the environment and the future were intimately intertwined. Futures research was on the rise both in Sweden and internationally. Many politicians, business leaders, scientists, and intellectuals had high hopes of the field. Others worried about the direction it would take. In May 1971, Prime Minister Olof Palme appointed a commission of enquiry to examine the possibilities of conducting futures studies in Sweden. The commission was led by Alva Myrdal, and her group included Martin Fehrm and Birgitta Odén. The group's report, *Att välja framtid* [English title: *To Choose a Future*], was submitted in the late summer of 1972 and led to the establishment of the Secretariat for Future Studies.[10]

The international debate about the future also made an impression in Sweden. Many international bestsellers, such as Alvin Toffler's *Future Shock* (1970) and Gordon Rattray Taylor's *The Doomsday*

---

9 Gösta Ehrensvärd, *Före – Efter: En diagnos* (Stockholm: Aldus, 1971).
10 Björn Wittrock, *Möjligheter och gränser: Framtidsstudier i politik och planering* (Stockholm: Liber förlag, 1980); Joar Tiberg, 'Vart tog framtiden vägen? Framtidsstudiernas uppgång och fall, 1950–1986', *Polhem: Tidskrift för teknikhistoria* 13.2 (1995), pp. 160–175; Jenny Andersson, 'Choosing Futures: Alva Myrdal and the Construction of Swedish Future Studies 1967–1972', *International Review of Social History* 51.2 (2006); Gustav Holmberg, 'Framtiden: Historikerna blickar framåt', pp. 281–288; Gustaf Johansson, *När man skär i nuet faller framtiden ut*, pp. 188–212; Carl Marklund, 'Double Loyalties? Small-State Solidarity and the Debates on New International Economic Order in Sweden during the Long 1970s', *Scandinavian Journal of History* 45.3 (2020); Karl Haikola, 'Historiska perspektiv på 1970-talet', *Scandia* 86.1 (2020).

Book (1970), were quickly translated into Swedish. But the most important publication was the Club of Rome's report *Limits to Growth* (1972). Published in March 1972, it was immediately cited in the Swedish debate about the future. What was special about the Club of Rome's predictions for the future was that they were based on computer simulations. Researchers used these to try to understand how variables such as raw-material resources, population growth, and environmental destruction interacted within a dynamic world system. The group's conclusion was that continued economic growth would lead to a global collapse.[11]

In the Swedish debate, however, the focus lay not on the Club of Rome's report but rather on Gösta Ehrensvärd's diagnosis. Later in this chapter I will therefore map out and analyse how that diagnosis circulated in the public sphere, from when *Före – Efter* was first published in October 1971 until the Stockholm Conference began in June 1972. First, though, I will examine the media storm around Hans Palmstierna in the late summer of 1971. What did he say that caused such an uproar? What were the reactions? And what position was Axel Iveroth really advocating?

Hans Palmstierna's speech

At 10 a.m. on 20 August, Hans Palmstierna was welcomed onto the podium at the Swedish Factory Workers' Union congress. It was held at Folkets Hus in Stockholm, and many journalists were present. Palmstierna was well prepared and had carefully orchestrated his own appearance. He had sent a transcript of his speech in advance to the editorial department of the social democrat daily newspaper *Arbetet* in Malmö, which told its readers what he would say at the congress.[12] Besides, he had published a cultural article in *Dagens Nyheter* three days before, about factory

---

11 Donella Meadows, Dennis Meadows, Jørgen Randers, and William Behrens III, *The Limits to Growth: A Report for the Club of Rome's Project on the Predicament of Mankind* (New York: New American Library, 1972); Patrick Kupper, '"Weltuntergangs-Vision aus dem Computer": Zur Geschichte der Studie "Die Grenzen des Wachstums" von 1972', in Frank Uekötter and Jens Hohensee (eds), *Wird Kassandra heiser?: Die Geschichte falscher Ökoalarme* (Stuttgart: Steiner, 2004); Elke Seefried, 'Towards the Limits to Growth? The Book and its Reception in West Germany and Britain 1972–1973', *German Historical Institute London Bulletin* 33.1 (2011).

12 Anon., 'Palmstierna på Fabriks kongress: Stoppa dubbelspelet forskare – konsulter', *Arbt*, 20 August 1971.

work and working-environment problems. It was grotesque, he wrote, that after nine years of schooling, young people were made to stand beside a conveyor belt to perform 'man-eating hard graft at high speed'. What made things even worse was that factory environments were noisy, mechanically risky, and often toxic. Such conditions could not be allowed to continue in a rich and highly developed country like Sweden.[13]

Palmstierna began his speech by emphasizing that it would be 'very personally coloured'. It started out from a trip he had made that summer to an unnamed factory workshop club in central Sweden. There he had been told that the workplace had major problems with 'something called epoxy resins'. The workers developed eczema, skin damage, and eye problems. One man had become disabled and been forced to retire early. 'What happened at this workplace is nothing remarkable', Palmstierna asserted. Precisely the same problem had existed for a long time at the large company Asea's workshop in Västerås. 'In the end, Swedish workers refused to take these jobs and let foreigners have them', he said. But why did the companies not share their experiences among themselves? How did the Federation of Swedish Industries function? Did it not issue any warnings to its members? 'This sort of thing should not be happening', thundered Palmstierna.[14]

From epoxy resins Palmstierna moved on to PCBs, a group of environmentally hazardous substances which were the focus of much attention in the Swedish environmental debate in the early 1970s. Among other things, the National Environment Protection Board published a report showing that high levels of PCBs were present in all Baltic Sea fish. This was affecting sex hormones and risked causing sterility in both animals and humans. Palmstierna emphasized that 'the people who are working on the factory floor and standing in the PCB fumes' were the most vulnerable. The dangers had been known since the 1890s, yet it was not until 1971 that the Riksdag had passed a law restricting PCB usage. The new law had been prompted by the damage seen in the external environment, a damage which was perceived as an immediate menace. 'It was not passed to protect the workers – we must be aware of that', Palmstierna pointed out.[15]

---

13 Hans Palmstierna, 'Tempoarbete och undervisning', *DN*, 17 August 1971.
14 Hans Palmstierna, 'Hälsorisker i svensk kemisk industri', pp. 1–2, 452/2/14 (HP ARBARK).
15 *Ibid.*, pp. 2–4.

Staying with this perspective, Palmstierna asked: Why was not more value placed on people's lives and well-being? Why were toxic substances not banned in workplaces? Did serious damage have to happen to fish and birds before society would intervene with any degree of strength? These questions had been behind his investigative work for the Factory Workers' Union; but his research had also raised new ones, including the question of who was responsible for the current situation. It was profoundly problematic, Palmstierna felt, that the duty of investigating health hazards in industry had fallen on the unions. Ought it not to be the employers' 'self-evident obligation to ascertain the risks associated with plastics and all sorts of things before adding such components to the manufacturing process?' He informed the congress delegates that a requirement to this effect had now been added to the Social Democrats' environmental programme.[16]

Another issue that his investigation had raised for Palmstierna was industry's relationship with the world of research. Initially, he said, the process of extracting data from some researchers had been 'surprisingly sluggish'. Upon examining them more closely, he had realized that not all of them were 'independent experts'. Indeed, a number of the 'sluggish information providers' were consultants for companies within the relevant industry. He particularly singled out the industrial development company Incentive. It was owned by the Wallenberg family, the dominant power group in the world of Swedish business. Founded in the early 1960s, Incentive had built up a strong and wide-ranging network of contacts within the research world. The fact that for some years now Swedish researchers had not been obliged to report consultancy work to their universities made it difficult to know who could be trusted. 'If I find that an expert's statement contradicts common sense or is evasive', Palmstierna said, 'I have become used to trying to find out where this expert's true home base is, and what it is that he's defending.'[17]

Palmstierna also attacked consultancy work for being an expression of growing class differences. It proved that people were accepting such work on the basis of their own interests rather than in the national interest. This situation made it harder for worker-protection measures to function, because as long as it was the companies that were paying the researchers, union interests would hardly be

---

16 *Ibid.*, p. 5.
17 *Ibid.*, p. 6.

prioritized. Palmstierna called on the government to invest large resources in occupational-safety measures. However, he was careful to emphasize that such a move should not take place at the expense of protection for the external environment; '[y]ou can cut funding for roads or whatever, but not from that which protects human beings and their environment'. Palmstierna envisioned a national supervisory testing laboratory which would ensure that companies provided correct information about various hazards and kept one another informed. The supervisory laboratory would be run by independent researchers who did no consulting work.[18]

Palmstierna concluded his speech by highlighting additional class aspects of the problem. He stressed that the risks associated with new chemical substances were unevenly distributed. The people who were most affected were workers on low wages. What made matters even worse was that such individuals could not afford to eat food that was as expensive and high in protein as high-income people could. The protection that a good diet could provide hence did not help those who were most at risk. 'So there is an ugly class-differentiating mechanism in this area which we must combat', he said. Nonetheless, Palmstierna was still hopeful about the future. He was particularly pleased that many young researchers came from working-class families, and he hoped they would retain a sense of solidarity with the working class throughout their working lives. In the long run, this solidarity would enable society to remedy the dangers in the internal environment.[19]

The media storm around Hans Palmstierna

Palmstierna's speech on 20 August at the Swedish Factory Workers' Union congress immediately became national news. Television and radio reported on it and interviewed him. The summary of his speech sent out by the news agency Tidningarnas Telegrambyrå had a particular impact. The newswire said that Palmstierna had accused Swedish scientists of being 'corrupt'. He had not in fact said so in his speech, but variants of the newswire were reproduced all over the country, especially in the local press.[20] Journalists at the

---

18 Ibid., pp. 7–8.
19 Ibid., pp. 8–9.
20 Cf. Anon., 'Docent Palmstierna anklagar vetenskapsmän för korruption', *Helsingborgs Dagblad* (HD), 21 August 1971; Anon., 'Svenska vetenskapsmän skylls för korruption', *Hudiksvallstidningen* (HT), 21 August 1971; Anon.,

major newspapers immediately began contacting researchers and industry representatives for comment.

On 21 August all the major newspapers featured articles about Palmstierna's speech. *Svenska Dagbladet* reported that the government's environmental protection expert had claimed that Swedish researchers 'were often bought by industry'. The consultants tended to sweep 'discoveries about health risks at workplaces' under the carpet. The researchers contacted by the newspaper had reacted strongly to the attack. It was 'frivolous, unpleasant, and ridiculous!' proclaimed the front-page headline. Professor Axel Ahlmark emphasized that 'a person in Palmstierna's position should avoid making such an undifferentiated and misleading attack'. Associate Professor Åke Swenson stated that 'no serious researcher wants to appear in public in order to gain big headlines before possessing reliable evidence for his information'.[21] The main editorial criticized Palmstierna for 'flailing about' and attacking an entire profession. It did not benefit environmental efforts and was scarcely apt to increase confidence in the government-appointed Environmental Advisory Council, said the anonymous writer.[22]

Dagens Nyheter also focused on Palmstierna's attack on the world of research. 'Can the general public and colleagues trust a university researcher who does extra work for industry?' wondered the reporter. 'Are Hans Palmstierna's allegations fair?' Professor Sune Bergström, rector of the Karolinska Institute and chairman of Incentive, felt it was 'shocking to treat an issue that is of such importance to the country's economy in this manner'. A small country like Sweden, he said, was completely dependent on its few experts helping industry to develop new products. Did Palmstierna want to clip the wings of Sweden's entire industrial base? Other representatives of Incentive expressed similar arguments. Its CEO, Sten Gustavsson, stated that the company was not ashamed of its contacts with researchers. On the contrary, it was in the interest of all of society to bridge the gap between industry and academia. By contrast, Associate Professor Stig Tejning in Lund expressed a diverging opinion. He certainly did not believe that industry-linked researchers had any malicious intent when they withheld information. 'Usually it boils down to a

---

'Svenska vetenskapsmän låter sig korrumperas', *Skaraborgs Läns Tidning* (*SLT*), 21 August 1971.
21 Göran Licke, 'Forskare om Palmstierna: Angrepp ohemult, obehagligt och löjligt!', *SvD*, 21 August 1971.
22 Anon., 'Forskarna', *SvD*, 21 August 1971.

simple lack of understanding', he said. 'They are not involved in environmental protection but instead regard production as the only important thing. An outdated view, quite simply.' Personally, he could not imagine doing any consulting work for industry.[23]

However, several other researchers at Lund University were sharply critical of Palmstierna. The Lund vice-chancellor, Professor Sven Johansson, argued that the accusation of corruption was an absurd exaggeration. *Sydsvenska Dagbladet* reported that several professors 'were yet again able to state that Palmstierna is not a witness for the truth' and that he dealt 'carelessly with facts'.[24] In *Kvällsposten*, Professor Maths Berlin, himself a consulting researcher, said he refused to believe 'that someone knowingly and deliberately conceals scientific information'. Hans Gullberg, secretary of an ongoing government enquiry into the work environment, denied that the enquiry had difficulty obtaining information. He could hypothetically imagine that the problem might exist; but he pointed out that researchers could sometimes have good reasons to withhold information, such as a desire not to create unnecessary panic.[25]

Still, *Kvällsposten* did speak with one researcher who fully agreed with Hans Palmstierna's 'roundhouse punches': Björn Gillberg. He was reported to have jumped for joy at Palmstierna's manoeuvre. 'It's terrific that we agree', remarked Gillberg, 'now Valfrid Paulsson at the National Environment Protection Board must think we're *both* crazy…'. He informed the reporter that Palmstierna was absolutely correct in his suspicions, singling out the National Board for Technical Development as a particularly flagrant example. Most of the professors who sat on that research board and handed out public resources were consultants for various Wallenberg companies. There were also many professors with divided loyalties at the National Environment Protection Board. 'I, at least, regard such things as bordering on corruption', said Gillberg.[26]

Palmstierna's standpoint was also supported by *Aftonbladet*, probably as part of an orchestrated media strategy. As early as 21 August, the newspaper published an in-depth report on how

---

23 Bo Teglund, 'Toppforskare om Palmstierna: Vill han vingklippa hela industrin?', *DN*, 21 August 1971.
24 Hans Widing, 'Fränt angrepp på forskare: "Många är korrumperade"', *SDS*, 21 August 1971.
25 Ulf Johansson, 'Angripna forskare ger svar på tal: – Vi är inte fega!', *KvP*, 21 August 1971.
26 *Ibid.*

governmental research funds were being distributed. The object of the attack was the above-mentioned National Board for Technical Development (STU). The journalist Thomas Danielsson had analysed which professors sat on the board, what ties they had to industry, and what proportion of the allocated grants went to members of the board. The report showed that the majority of the board's members were linked to the Wallenberg sphere. One of them was Carl-Göran Hedén, professor of bacteriology at the Karolinska Institute. The autumn of 1967 had seen him in hot water because of his chapter in *Människans villkor*, where he argued that scientists should be given greater influence over political decision-making. *Aftonbladet*'s report revealed that, in both 1970 and 1971, Hedén and his research team had received about half of the grants that STU had to distribute. The reporter found this behaviour – as well as the ties to Incentive – to be severely compromising.

Aftonbladet announced that it had spoken with many researchers who underlined STU's great power. The board members were the ones who decided what would be researched in the biotechnology field and who would do it. One anonymous researcher said that 'a researcher who publicly criticizes or sounds the alarm can quite quickly count on reprisals from STU'. These might occur in the form of reduced or no grants, which meant that the researcher would be forced to apply for a new job. 'This effectively silences you', said the anonymous researcher.[27] It is no wild guess that the person the reporter spoke to was Björn Gillberg.

Clearly, then, the immediate press reactions adhered to ideologically predictable lines. Palmstierna was supported by the working-class press, whereas the non-socialist press rallied behind industry and the research community. This pattern becomes even more apparent from a review of the local press. For example, the social democratic newspaper *Värmlands folkblad* rejoiced that someone of Palmstierna's calibre had finally banged his fist on the table. 'When he sounds the alarm, people *usually* listen.'[28] In contrast, the conservative *Norrbottenskuriren* was more sceptical: 'You have to be a very distrustful Socialist to have such a poor opinion of your fellow human beings', wrote the newspaper. 'Compared to the government,

---

27 Thomas Danielsson, 'Så fördelas svenska statens forskningspengar av storkapitalets experter', *AB*, 21 August 1971.
28 T. B., 'Palmstierna slår larm igen', *Värmlands folkblad* (*Vf*), 21 August 1971.

The counterattack on Hans Palmstierna

The media storm around Hans Palmstierna entered a new phase on 25 August, when the tabloid *Expressen* published a long article entitled 'Vår miljövårds väckelsepredikant' [The environmental revivalist preacher of our time]. It was a direct counterattack on Palmstierna. What did his own loyalty ties look like? Why had he acted the way he did? The anonymous writer began with an account of Palmstierna's sweeping criticism of his 'corrupt and morally reprehensible' colleagues. He had alleged that researchers who did consultancy work for industry were driven by personal greed and would not 'lift a finger to help workers escape from an unhealthy environment'. However, Palmstierna had not presented any concrete evidence that this was so. Yet the writer felt there was something 'a bit sweet' about the agitated attack. In a follow-up TV interview, Palmstierna had shown that he understood the situation of the denounced researchers. 'Of course, a person is not eager to attack someone who gives them a large proportion of the honey on their slice of bread.'

In *Expressen*, this statement was turned against Palmstierna himself. The course of his life was outlined with quick strokes of the pen. He had been a Social Democrat for a long time, and in the 1960s he had become 'a minor celebrity in the party'. But he had been considered as being 'outside the mainstream and a little hard to place'. His political career had never gained any real momentum. 'But in 1967 he became a big celebrity when he published the little manifesto *Plundring, svält, förgiftning*.' It had launched the environmental debate and made Palmstierna one of the country's most sought-after speakers. 'Because he belonged to the right party', the writer pointed out acidly, 'of course he became an expert in government circles.' In subsequent years Palmstierna's name had been on Social Democratic lists for Riksdag members, but never in a really electable place. It was rumoured that he was now eyeing a ministerial post. 'In such circumstances, it is naturally reasonable that a person would try to express himself in the right way when giving a speech at the Factory Workers' Union's congress.' It could not hurt to keep well in with those who buttered your bread.

---

29 Anon., 'Palmstierna anklagar', *Norrbottenkuriren* (NK), 21 August 1971.

The counterattack on Palmstierna did not stop there. The writer also questioned his view of humanity. Did Palmstierna really believe that researchers with a working-class background displayed greater solidarity and were more ethical than others? Was this not, in fact, an expression of the 'social-group mystique' that had dominated the 'old swamp-like class society'? That society had believed that the children of 'better folk' commanded greater intellectual and ethical powers than others. According to such a view, the best thing working-class children could do was not to throw their weight about. 'That way, everyone would remain in their appointed role and society would continue to prosper.' It was all 'quite cynical' and 'of course completely preposterous'. But now, in 1971, here comes Palmstierna and says the same thing as those self-satisfied upper-class bigwigs of the 1890s – except in reverse.

And who was Palmstierna to make such claims? Was he not himself from 'an old prestigious noble family that had been promoted to baronial status back in the time of King Fredrik I of blessed memory [that is, in the eighteenth century]?' Nor could his living conditions be characterized as basic. It was said that he lived in a detached house in a Stockholm suburb and had a reported income that was about four times as high as what 'an ethically honourable working-class family has to manage on'. What was more, he drove a car to work instead of travelling by public transport. Should he be doing that? Cars spread toxic exhaust fumes. Were cars not more dangerous to the environment than the occasional associate professor doing a bit of work on the side for industry?[30]

This caustic onslaught on Palmstierna in *Expressen* was not an isolated phenomenon. That same day, the liberal paper *Göteborgs-Posten* characterized him as a 'sideways-promoted combatant'. The anonymous editorial writer said that Palmstierna's rise through the Social Democratic ranks had not left a single mark on the government's environmental policy. On the contrary, a series of political decisions had 'directly contradicted the ideas that Hans Palmstierna had championed before his elevation'. True, the Social Democrats' environmental programme was imbued with fine words and big ambitions – but they were hardly being translated into action. At the environmental policy level, Palmstierna was now on track to 'amass as many defeats as he had for a while amassed victories'.[31]

---

30 Anon., 'Vår miljövårds väckelsepredikant', *Exp*, 25 August 1971.
31 Anon., 'Bortbefordrad kämpe', *GP*, 25 August 1971.

In the popular weekly magazine *Se*, hard-hitting journalist Rune Moberg made a sneering personal attack in the spirit of *Expressen*. He said he felt a bit sorry for 'upper-class types who want to help fight for the workers' cause'. They had problems finding the right tone and often exerted themselves so much that their voices cracked. 'Recently we saw a real live baron talking with burning intensity about the forgotten men on the workshop floor.' Did Palmstierna want to emulate Olof Palme in more than just having a similar name? What was this 'scientist from the high nobility' really thinking when he accused his research colleagues of withholding information 'for the sake of filthy lucre'?[32]

The media hullabaloo reached a climax with Axel Iveroth's speech in Jönköping on 1 September. The chairman of the Federation of Swedish Industries found it most remarkable that Palmstierna sought to 'clothe environmental-protection work in outdated terms of class struggle'. His attack on scientists with industry links was 'sickening' and 'malicious'. It proved that Palmstierna was no longer an environmental alarm clock. With him as secretary of the Environmental Advisory Council, there were no guarantees that a healthy climate of cooperation could evolve. Such a climate, Iveroth stressed, was necessary in order to deal with the problems.[33]

Iveroth added that it was difficult 'to appear as a representative of industry in today's environmental debate'. It was a deeply thankless role because 'irate critics' wanted to create conflicts almost daily. 'They seek to gain publicity and believe they are reaping political laurels through dogmatic, clamorous, and inexpert attacks on industry and its environmental work.' Nor did they offer any concrete alternatives. Such critics made it 'indecently easy' for themselves by keeping silent about the consequences that various measures would have on employment, living standards, and material prosperity in the country. At the same time, industry had reason to be proud of its history. 'There is no doubt', Iveroth asserted, 'that industry's environmental-protection ambitions awoke earlier than the general environmental awareness.' Companies had been seeking to identify and solve problems affecting both the internal and the external environment for decades. In actual fact, said Iveroth, industry had become aware of the environmental problems 'decades before the politicians and responsible authorities – let

---

32 Rune Moberg, 'Blinka lilla Stierna där', *Se*, 36 (1971).
33 Iveroth, 'Industrin och miljövården', pp. 6–7.

alone the press – had begun to realize the scope and importance of the field'.[34]

Iveroth pointed out that the water-management committee of the Swedish Forest Industries Federation would soon be thirty years old, and that the Swedish cement industry's investments in air protection had totalled many millions of kronor long before the Environmental Protection Act had come into force. He emphasized that in the early 1960s the Federation of Swedish Industries had initiated the founding of a semi-public-, semi-private-sector institute for research into water and air protection.[35] The institute had been set up at a point in time when there was no government counterpart to negotiate with. Furthermore, said Iveroth, Swedish industry's total emission volumes 'are in fact steadily declining already'. He underlined that 'companies everywhere are pursuing solutions that are both acceptable from an environmental point of view and financially reasonable'. Many major companies – such as the mining company Boliden and the food company Felix – were working hard to reduce and minimize their environmental impact. They were not waiting for political decisions and new environmental legislation. Indeed, business companies were leading the way.[36]

The politicization of environmental issues

What is clear from Axel Iveroth's speech, as well as from the personal attacks on Hans Palmstierna in the press, is that in the early 1970s it was not only the Social Democrats who were trying to advance their positions and turn the environmental debate in a new direction. Swedish industry and the non-socialist press were attempting to mobilize, too. Iveroth and Palmstierna applied contradictory approaches. Nature and Youth Sweden and other environmental movements were offering further options. The established party that best succeeded in capturing the growing environmental commitment was the Centre Party. In the 1970 election, the party received 19.9 per cent of the vote. In 1973 it achieved 25.1 per cent. The Centre Party's version of the environmental

---

34 Ibid., pp. 2–3.
35 For studies of this type of research institute, see Ingemar Pettersson, *Handslaget: Svensk industriell forskningspolitik 1940–1980* (Stockholm: KTH Royal Institute of Technology, 2012).
36 Iveroth, 'Industrin och miljövården', pp. 3–4.

debate focused on regional policies and on strengthening local communities.[37]

Consequently, the politicization of environmental issues was a process that was both conflict-filled and multifaceted. It was also intimately connected with an important stage in the media history of the sciences.[38] Around 1970, new forms of scientific activism gained ground. Many researchers spoke out in the media in ways that contrasted with the traditional ideal of researchers as distant and apolitical beings. True, public-speaking and politically committed scientists were not a new phenomenon; but during this period their numbers grew considerably. Many of them also collaborated closely with the new social movements. A direct consequence of this was that scientific controversies were increasingly often conducted in public.[39]

Many members of the world of research were sceptical about this development. This scepticism comes out not least in the reactions to Hans Palmstierna's speech. In the words of Associate Professor Åke Swenson, Palmstierna did not act like 'a serious researcher'. After all, such a person would not use the media to gain political or career benefits. Biologist Ingvar Wiberger asserted that Sweden had begun 'to teem with playboy researchers', and he singled out 'the environmental gods' Hans Palmstierna and Björn Gillberg for special mention. 'It is ridiculously easy to climb the career ladder as an "environmental yeller" today', he said. 'It's much harder to do it as a meticulous researcher.'[40]

One high-profile figure who agreed with the critics was Professor Gunnar Hambraeus, president of the Royal Swedish Academy of

---

37 Holmberg, *Längtan till landet*.
38 Ekström, 'Vetenskaperna, medierna, publikerna'.
39 Rae Goodell, *The Visible Scientists* (Boston, MA: Little, Brown, 1977); Bess, *The Light-green Society*, pp. 76–79; Stephen Bocking, *Nature's Experts: Science, Politics, and the Environment* (New Brunswick: Rutgers University Press, 2004), pp. 55–59; Egan, *Barry Commoner and the Science of Survival*, pp. 9–11; Jon Agar, 'What Happened in the Sixties?', *British Journal for the History of Science* 41.4 (2008), pp. 573–574.
40 Rune Gustafsson, 'Biolog vid Bofors går till attack mot "Playboy-forskare"', *Karlstads-Tidningen* (*KT*), 4 September 1971. In September and October the interview was published in several other local newspapers, such as *Norrköpings Tidningar-Östergötlands Dagblad* (*NT-ÖD*), 27 September 1971, *Jönköpingsposten* (*JP*), 28 September 1971, *Norra Skåne* (*NS*), 28 September 1971, *Ljungbytidningen* (*LT*), 29 September 1971, *Borås tidning* (*BT*), 11 October 1971, and *HT*, 11 October 1971.

Engineering Sciences. On 22 October 1971, he gave an inaugural speech at an environmental trade fair in Gothenburg. His speech attacked 'the doomsday prophets' Hans Palmstierna and Björn Gillberg. Hambraeus said that their sole achievement was to have created hysteria and frightened people. 'Before, people were afraid of hell. Now we're afraid for the environment. That is a regrettable result of about 100 years of industrialization efforts.'[41]

However, the debate among researchers was not completely polarized: some attempted to assume a mediating and reasoning position. One of those was Birgitta Odén, who felt that the debate about the researchers' loyalties had been distorted in an unfortunate manner. 'Palmstierna's demagogic simplifications from the rostrum have warped what is fundamentally a social problem into an ethical problem about individuals', she wrote. In addition, it was obvious that Palmstierna's attack on the research world and on industry constituted an 'impermissible generalization'. But the generalization offered by the other side was no more reasonable. In her experience, researchers were neither more nor less ethical than other groups. She felt that the really interesting question was what the research structure in Swedish society should in fact look like.

Odén stressed that she had no objections in principle to doing research on contract. In a small country like Sweden, with very limited human resources, contacts between industry and academia could not be dispensed with. Nor was industry the sole beneficiary from such contacts. Doing contract work stimulated university researchers to find new research proposals and areas of application. However, there was an immediate danger that industrial researchers would merely be used to develop new products. That was certainly important, but it was not the only thing that qualified researchers should devote their time to. It was equally important that there should be researchers working to solve the undesired problems that arose from modern industrial society. Seen from this point of view, contract research did pose an indirect threat. 'Society's resources for utilizing the top expertise for problem-orientated research', Odén wrote, 'appear very small in comparison with those of industry.' She called for another type of contract research: one that sought to create a better world for everyone. In this context, the client was neither industry nor the trade-union movement. It was 'the anonymous

---

41 Elisabeth Wiechel, 'Gillberg-Palmstierna kritiseras: "De skapar miljövårdshysteri"', *GP*, 23 October 1971.

fellow human being' who should be given a 'totally different and more recognized place in our research structure'.[42]

Odén's vision of problem-orientated and socially beneficial research had been developing for a long time. She had been promoting it in scholarly contexts since the late 1960s.[43] In the autumn of 1971, however, she seized the opportunity and went public. Just over a week after her first article she developed her thoughts further, beginning with the almost unanimous condemnation with which the world of research had greeted Hans Palmstierna's speech. Odén felt that it was an illustrative example of how the world of research exercised control over dissent. There were three deadly sins in the world of research, she said: being political, being wrong, and demonstrating poor judgement. Palmstierna had committed all of them. His actions directly contradicted the research world's fundamental values – and that was why he had been rejected as a matter of instinct.

Birgitta Odén pointed out that, as a historian, she had no opinion about Palmstierna's qualities as a scientist. However, she was personally convinced that he was a 'politically deeply committed individual'. Clearly he had sometimes been wrong and at other times demonstrated a lack of judgement. 'No one who has shouldered tasks of the format chosen by Hans Palmstierna can avoid falling into the research world's three deadly sins', she asserted. But did that make it self-evident that his warnings should be rejected?

In her view it did not. She argued that the world of research, 'made wise by the many transgressions against dissidents committed in the history of science', should think twice. Should not someone like Hans Palmstierna who, despite his personal shortcomings, had previously shown himself 'to be so fundamentally right' be worth listening to extra carefully? Perhaps he was 'fundamentally right about' something this time too, despite the 'demagogic simplifications'? Odén argued that the fear of being excommunicated by the world of research was dangerous. It threatened to undermine the critical social role of research. Because who would wake society up if the scientific alarm clocks were silenced? 'The Federation of Swedish Industries?'[44]

---

42 Birgitta Odén, 'Samhället måste bygga upp kontakten med forskarvärlden', *Arbt*, 11 September 1971.
43 Birgitta Odén, 'Clio mellan stolarna', *Historisk tidskrift* 88 (1968); Birgitta Odén, 'Historiens plats i samfundsforskningen', *Statsvetenskaplig tidskrift* 71.1 (1968).
44 Birgitta Odén, 'Forskare får inte tysta kritiska väckarklockor', *Arbt*, 19 September 1971.

Odén's contribution to the debate did not make much of an impression, even though *Arbetet*'s editor repeatedly contacted Hans Palmstierna asking for a reply.[45] But her argument was supported by the group, led by Alva Myrdal, which was investigating the possibilities of Swedish future studies. The group's express ambition was that significant public research resources should be invested in tackling society's major challenges. Later in the 1970s, ambitious interdisciplinary projects were launched in order to study the future energy supply and security situation.[46] When Linköping University was founded in 1975, problem-orientated, socially beneficial research was a guiding principle for its activities. The level of conflict in the public debate was one thing; practical research policy was another.

Yet the two spheres were connected. As we have seen, many actors – such as Hans Palmstierna, Birgitta Odén, Valfrid Paulsson, and Carl-Göran Hedén – moved between them. There were also actors who, without appearing in the public eye themselves, sought to advance the debate. One of them was Daniel Hjorth, head of publishing at Aldus. The publishing firm was owned by Bonniers, the dominant media group of companies in Sweden, and had been publishing popular-science paperbacks since the early 1960s.[47] On 3 September 1971, Daniel Hjorth wrote to Hans Palmstierna asking if he would like to develop his thoughts about researchers' dependence on industry in a debate book.[48] Indeed, Palmstierna would have liked to, but he was already under contract. In December 1972, the book *Besinning* [approx. Coming to our senses] was released by another publisher.[49] By that time, though, Aldus had already achieved other successes. They began with the publication in October 1971 of Gösta Ehrensvärd's *Före – Efter: En diagnos* [Before – after: a

---

45 Two cards from Levi Svenningsson to Hans Palmstierna, undated in September 1971, 452/2/14 (HP ARBARK).
46 Andersson, 'Choosing Futures'; Karl Haikola, 'Objects, Interpretants, and Public Knowledge: The Media Reception of a Swedish Future Study', in Östling, Larsson Heidenblad, and Nilsson Hammar (eds), *Forms of Knowledge*.
47 Ragni Svensson, 'Pocketboken gjorde kunskapen till en konsumtionsvara', *Respons* 2 (2020).
48 Letter from Daniel Hjorth to Hans Palmstierna, 3 September 1971, 452/2/14 (HP ARBARK). For Aldus, see Per Gedin, *Förläggarliv* (Stockholm: Bonnier, 1999).
49 Hans Palmstierna with assistance from Lena Palmstierna, *Besinning* (Stockholm: Rabén & Sjögren, 1972).

diagnosis], whose circulation in the public sphere will now be examined in some detail.[50]

The diagnosis emerges

Swedish media began to report on Ehrensvärd's predictions for the future in mid-November 1971. At that time, media interest was solely manifested in the form of reports and reviews, which often simply described the content of the book.[51] The writers consistently emphasized that the author's vision of the future was no unsupported fantasy but was based on hard facts and mathematical calculations. However, *Svenska Dagbladet*'s Tom Selander stressed that much of Ehrensvärd's diagnosis could not be 'other than qualified guesses', albeit 'based on facts'.[52] Eva Moberg made a similar reservation in her review, writing: 'Although the prediction does not appear at all unreasonable, it is still in some way unrealistic not to leave any room for the unpredictable. We do know that the unpredictable will happen, even though we do not know what it is.'[53] There were no more critical comments than these in 1971, and reviewers did not doubt that Ehrensvärd's vision of the future constituted new and urgently necessary knowledge. An editorial by the monthly cultural magazine *Vi*, founded by the Swedish Cooperative Union in 1913, said it was 'one of the autumn's most important books',[54] and in *Norrköpings Tidningar*, a local newspaper with a non-socialist orientation, Bengt Sjönander wrote that it was a 'deeply shocking book that every single person should read a bit of every day for years'.[55]

---

50 The following is based on Larsson Heidenblad, 'Framtidskunskap i cirkulation'.
51 Göran Larsson, 'Hungerkatastrof för oss tillbaka till medeltiden', *AB*, 10 November 1971; Tom Selander, 'Den mörknande framtid', *SvD*, 11 November 1971; Bertil Walldén, 'Gösta Ehrensvärds syn på framtiden', *Vlt*, 16 November 1971; Bengt Hubendick, 'Människan på jorden', *GHT*, 25 November 1971; Eva Moberg, 'Spara något åt barnbarnsbarnen', *Vi*, 27 November 1971; Josef Lövgren, 'Våra barnbarns barn kannibaler?', *Gefle Dagblad (GD)*, 27 November 1971; Lars Gyllensten, 'Industrisamhällets bokslut', *DN*, 5 December 1971; Sven Rinman, 'Mot katastrofen', *GHT*, 11 December 1971; Bengt Sjönander, 'Ett perspektiv på framtiden', *NT-ÖD*, 18 December 1971; Allan Fredriksson, '10 miljarder svälter ihjäl – och det är bara hundra år dit!', *Göteborgs-Tidningen (GT)*, 20 December 1971.
52 Selander, 'Den mörknande framtid'.
53 Moberg, 'Spara något åt barnbarnsbarnen'.
54 Sten Lundgren, 'Bara ett liv', *Vi*, 27 November 1971.
55 Sjönander, 'Ett perspektiv på framtiden'.

It was also telling that Bengt Hubendick, one of the most prominent environmental voices in Sweden in the years around 1970 and a frequent contributor to the Gothenburg-based and liberal-orientated broadsheet *Göteborgs handels- och sjöfartstidning*, highlighted Ehrensvärd's diagnosis early on in order to reinforce his own message about the unsustainability of modern industrial civilization. In his review, Hubendick wrote that the present age was 'an anomaly, an abnormal situation, a degenerate episode in human history. Our direction of development is heading towards an abyss. Yet we are constantly told that development must run its course. Nonetheless, we are rushing, even racing, one another towards the abyss.'[56] Hubendick's scathing critique of society foreshadowed the polarized debate about the future that would flare up in February 1972.

Another article from 1971 that was a harbinger of the future was Lars Gyllensten's review of the book in *Dagens Nyheter*. Gyllensten, a famous author and a member of the Swedish Academy, began with a historical review of doomsday prophecies and singled out the legend of the Tower of Babel as 'one of the foremost archetypes in our mythological equipment with which we and our ancestors have tried to interpret our destinies and adventures'. According to Gyllensten, the historical experience of unfulfilled doomsday prophecies was a dilemma for modern people because 'this circumstance in itself has a soporific effect – it is easy for us to shrug our shoulders when new ominous signs are cited about what will happen: we have heard similar prophets before, and things often turned out better than the fears predicted'. Against this background, he was careful to emphasize Ehrensvärd's 'high credibility' and 'solid calculations'.[57] It is clear from his review that Gyllensten himself took Ehrensvärd's diagnosis seriously while being well aware that a historical critique of it would be an obvious first rejoinder.

In a later phase of the book's circulation in the public sphere, Gyllensten's dilemma would arise on a broad front; but in 1971, no exchanges of opinion about Ehrensvärd's diagnosis occurred. Nor were there any interviews with its author, at least none that have been preserved for posterity.[58] Ehrensvärd's forecast was thus

---

56 Hubendick, 'Människan på jorden'.
57 Gyllensten, 'Industrisamhällets bokslut'.
58 According to Josef Lövgren's review 'Våra barnbarns barn kannibaler?' in *GD*, 27 November 1971, TV2's current-affairs talk show *Kvällsöppet* did interview Ehrensvärd. Unfortunately, according to both the SMDB database

circulating in the media during this initial phase, but it had not yet become a focus of public debate.

It is worth noting that towards the end of 1971, environmental issues were manifested in popular cultural form in the film *Äppelkriget* [The Apple War] by comedy duo Hasse Alfredson and Tage Danielsson. It depicted how foreign exploiters and unscrupulous politicians threatened a rural idyll. The film's theme song, 'Änglamark' [approx. A soil blessed by angels], was written by Evert Taube, a Swedish composer of songs and a troubadour with national-treasure status (he is depicted on the current 50-SEK banknote). Both the film and the song became classics, and today they are both closely associated with the environmental engagement of the 1970s. There and then, however, other historical actors made the biggest impact in the public sphere.

The breakthrough of the diagnosis

At the beginning of January 1972, media interest in Gösta Ehrensvärd's book intensified when several of the country's editorial pages used it as a discussion point to begin the new year.[59] The broadsheet *Skånska Dagbladet*, a Centre Party newspaper widely read in Skåne, especially in its rural parts, wrote that the book gave a 'frightening and shocking picture of what awaits human beings on earth', maintaining that 'the feeling that we are living in a manner that is hostile to human life in the long term' was spreading.[60] Social democrat broadsheet *Arbetet* stressed that 'what is happening now cannot be compared with anything that has happened before in human history',[61] and the middlebrow weekly magazine *Vecko-Journalen* described the book as 'dreadful in the true sense of the

---

and the National Library of Sweden's research service, no *Kvällsöppet* from November 1971 has been preserved.
59 Mario Grut, 'Åter till naturen!', *AB*, 2 January 1972; Gunnar Fredriksson, 'Vänstern behöver ditt förnuft', *AB*, 15 January 1972; Janerik Larsson, 'Kan människosläktet överleva år 2000?', *SDS*, 2 January 1972; Anon., 'Ny stenålder?', *SkD*, 3 January 1972; Anon., 'På lång sikt', *Arbt*, 3 January 1972; Gustaf von Platen, 'Bokslut över våra tillgångar', *Veckojournalen* (*VJ*), 4 January 1972; Hans Rudberg, 'Optimisten och pessimisten', *DN*, 11 November 1972; Sander, 'Vägen till helvetet', *DN*, 23 January 1972; Anon., 'Lyssnar någon?', *Land*, 14 January 1972; Anon., 'Den nödvändiga kompromissen', *Land*, 21 January 1972.
60 Anon., 'Ny stenålder?'
61 Anon., 'På lång sikt'.

word – worthy of provoking dread'.[62] At this time, the media also began to comment on the growing interest in Ehrensvärd's diagnosis, and the social democratic tabloid *Aftonbladet* published the first in-depth critique of the book.

The author of the critique was the writer on cultural matters Mario Grut, who placed Ehrensvärd's prediction within a long tradition of doctrines of doom, such as the Book of Revelation and Oswald Spengler's *The Decline of the West* (1918–1922). Arguing that this tradition was 'more philosophical than scientific', Grut focused on Ehrensvärd's ideological and political foundations. Grut labelled the Lund professor a reactionary for hoping that in the centuries following the great collapse of society, the technological expertise of the current period could be preserved in small scientific enclaves: 'Friedrich Nietzsche, the German Superman philosopher, would have smiled in recognition', Grut wrote, and he went on to criticize Ehrensvärd for his not only elitist but also Eurocentric points of departure.[63]

A related critique was expressed a few days later by the philosopher Paul Lindblom in *Arbetet*. He maintained that research into the future was currently fashionable, but that the interesting thing about it was not the predictions themselves – 'they won't come true' – but rather the examination of the underlying values. 'These values are often hidden', Lindblom wrote; 'people assume, for example, that American capitalism has given rise to a way of life that should be preserved without directly accounting for this dubious premise. But this value governs the prognoses to some extent.'[64] Lindblom thus assumed a sceptical attitude to Ehrensvärd's failure (in his view) to be open about his own ideological premises, arguing that this failure had consequences for Ehrensvärd's credibility as a scientist. Grut's review said that Ehrensvärd had 'moved into politics through the back door', a situation that called for stringent ideological analysis.[65] It is, however, important to observe that these two critical articles from the political left were not countered by Ehrensvärd personally, nor were they picked up by other debaters to any noticeable extent.[66] Nor did other writers follow suit.

---

62 Von Platen, 'Bokslut över våra tillgångar'.
63 Grut, 'Åter till naturen!'
64 Paul Lindblom, 'Vi vet inte mycket om framtiden', *Arbt*, 7 January 1972.
65 Grut, 'Åter till naturen!'
66 The exception is Karin Johansson, 'Måste vi spara nu?', *SDS*, 20 January 1972, which defends Ehrensvärd's diagnosis.

Left-wing criticism did not gain a foothold in the media circulation around the book.

The rapidly growing interest in Ehrensvärd's diagnosis was most clearly expressed on Sunday 9 January, when the front page of *Dagens Nyheter* ran a full-length picture of him. Under the headline 'Goodbye to prosperity, now we trust in the sun', it was said that humanity would never again be able 'to experience such prosperity as in the 1960s and 1970s'. Inside the newspaper was the first interview with Ehrensvärd in the Swedish media. The reporter stressed that for anyone who trusted his predictions, it was 'madness to keep increasing production and consumption'. The article gave considerable space to how Ehrensvärd himself believed the situation should be handled, highlighting his belief that the UN should 'place the whole world in a state of emergency'. At the national level, he thought that politicians should call a halt to production and introduce ration cards. Regretting that the serious situation required such 'dictatorial methods', Ehrensvärd emphasized that 'now, or definitely in the 21st century, the crisis programme will in any case be necessary in some way'.[67] The interview's hands-on orientation indicated that Ehrensvärd's diagnosis was moving ever closer to the sphere of current politics. A week later, *Dagens Nyheter* followed up the article by interviewing Prime Minister Olof Palme and the Liberal [People's] Party leader Gunnar Helén about how they regarded Ehrensvärd's vision of the future.

The attitude of the two politicians was summarized on the paper's front page by the headline 'We do not believe in doom'. The lead paragraph stated that humanity's standard of living could continue to rise, and no one needed to 'fear that we are currently wasting so much that there will be nothing left for our children and grandchildren'. Palme and Helén possessed 'a strong belief in continued technological progress' and asserted that 'when coal and oil run out, we will find new energy solutions'. This was contrasted with Gösta Ehrensvärd's predictions of 'the rapid downfall of Western society in a severe supply disaster'. The front-page presentation was highly polarized and had apocalyptic overtones. However, the longer interviews inside the newspaper presented a different picture.

Palme asserted that theorists of the future could be sorted into two schools: 'One is the happy technocrats – of whom I feel very distrustful – and the other is the one that keeps talking about the catastrophe. The danger with the latter is that we could become desensitized. That must

---

67 Björn Berglund, 'Uran, kol, olja – allt sinar', *DN*, 9 January 1972.

not happen.' Instead, the prime minister argued that there were great possibilities of remedying the serious situation by political means: 'I do not believe, as Ehrensvärd does, that the catastrophe will come, but I fully share his demands for political action.'

The interview with Gunnar Helén also differed in character from the front-page headline and lead paragraph. The Liberal leader said that it was 'possible that we are heading for a disaster', and that the next few decades will be 'a race against time for democracy as a system, for a solution to the population issue, for the supply of food and energy'. He underlined that Ehrensvärd's theories were 'within the bounds of possibility', and he reacted strongly when the reporter asserted that humanity had coped with prophesied disasters before: 'What *has* happened can never be proof of what will happen. So that line of reasoning is of no interest', he said. Helén thus did not distance himself from Ehrensvärd; if anything, he tended to side with him. However, he and Palme did stress that they believed there were political solutions that could be implemented within the prevailing democratic system.[68] Nevertheless, the polarized headlines – which magnified the dividing lines between Ehrensvärd and the politicians – are interesting in themselves and show how the media seized on conflicts and disagreements. That orientation on the part of the media was to be particularly characteristic of the third phase in the circulation of Ehrensvärd's diagnosis.

In January 1972 the Swedish debate about the future was still in its infancy, even though the media attention surrounding Gösta Ehrensvärd's diagnosis was both extensive and growing. Informative reviews and descriptions of the book continued to be published[69] as well as interviews with and about Ehrensvärd.[70] *Skånska Dagbladet* described him as a 'pessimistic professor';[71] and in a double-page spread in the popular financial magazine *Veckans affärer*, his diagnosis

---

68 Björn Berglund and Carina Fredén, 'Kan ni rädda oss?', *DN*, 16 January 1972.
69 Björn Nihlén, 'Det börjar 2050: Först global hungerkatastrof. Sen primitiv jordbrukskultur', *SkD*, 14 January 1972; Karl-Erik Eliasson, 'Stoppa rovdriften och befolkningsexplosionen', *HD*, 23 January 1972; Anon., 'Två svenska profeter', *UNT*, 26 January 1972.
70 Barbro Josephson, 'Sammanbrott kan undvikas', *DN*, 22 January 1972; Oscar Hedlund, 'Det är hans jobb att se in i framtiden', *DN*, 23 January 1972; Anon., 'Framtidsforskaren: Anpassningen måste börja nu – *innan* resurserna tar slut', *Veckans Affärer (VA)*, 27 January 1972; Birgit Lusch-Olsson, 'Återgång till det jordnära?', *SkD*, 28 January 1972.
71 Lusch-Olsson, 'Återgång till det jordnära?'

was referred to as 'a new doomsday prophecy'. The magazine also asked Ehrensvärd straight out if he was a pessimist about the future. He replied rhetorically: 'Is it pessimism to look forward to an agrarian society after years of want and decay, a society with an admittedly low but still secure standard? Is a simplified life so frightening to industrialized human beings with all their social prestige?'[72] Ehrensvärd was clearly not comfortable about having begun to be depicted as a doomsday prophet, and a few weeks later he sharply criticized the media image of him as 'somewhat vulgar propaganda'.[73] It was also clear from the introduction to *Före – Efter* that Ehrensvärd believed he was 'doing a balancing act between optimism and pessimism'. He asserted that he was 'deeply pessimistic about the short-term perspective', but said that he was simultaneously nourishing an 'unquenchable optimism about human tenacity and resilience in trying circumstances, far into the future'.[74] Ehrensvärd took pains to convey this complex picture in interview situations, but the media coverage – especially at the headline and lead-paragraph levels – allowed no room for such nuances.

Towards the end of January 1972, it was also obvious that Ehrensvärd's book had become a bestseller: it had sold 30,000 copies and now appeared in a fifth edition. The book topped *Vecko-Journalen*'s list of bestselling books nationwide. Daniel Hjorth at Aldus commented on the phenomenon in *Sydsvenska Dagbladet*: 'In Aldus's history, we have never had a book that was such a swiftly accelerating success. It is selling better now than before Christmas, and demand is only growing.'[75] What Hjorth probably also suspected at that point was that the attention being paid to the book and its forecast would soon increase even further.

The diagnosis is challenged

On 4 February, Aldus published a new book about the future: nuclear physicist Tor Ragnar Gerholm's *Futurum exaktum:*

---

72 Anon., 'Framtidsforskaren: Anpassningen måste börja nu'.
73 Anon., 'Ehrensvärd vidhåller: Vi måste återgå till jordbrukssamhället', *UNT*, 3 February 1972.
74 Ehrensvärd, *Före – Efter*, p. 6.
75 N. G. N., 'Den mörknande framtid', *SDS*, 28 January 1972. The sales successes are also mentioned in Bernicus, 'Svepet', *GP*, 30 January 1972; Janerik Larsson, 'Hur ser framtiden ut?', *SDS*, 30 January 1972; Advertisement in *DN* for *Före – Efter: En diagnos*, 4 February 1972.

Fortsatt teknisk utveckling? Spekulationer om problem som måste lösas före år 2000 [The 'future perfect': continued technological development? Speculations about problems that must be solved before the year 2000] (1972). Gerholm argued that the debate about the future had derailed because constructive confidence had been overshadowed by pitch-black cultural pessimism. He especially attacked radical, ecologically orientated criticism of society and the contemporary world: 'We are now told that the blessing of industrialization is nothing but hollow lies. It is confidently proclaimed that the welfare society's glimpse of prosperity is a crazy episode in the history of humanity. We will soon have emptied the Earth's storehouse of natural resources, and therefore we will be forced back into agrarian society's grey drabness and threadbare destitution.'[76] Gerholm argued that this gloomy vision of the future was to a great extent unjustified and also dangerous, because it could lead to a social 'paralysis precluding action, a paralysis that turns pessimism into a self-fulfilling prophecy'.[77] He believed that the answer to the challenges facing humanity was not to slow down development but rather to strive for new technological, scientific, and economic gains. If that happened, there was 'good reason to hope for a completely natural and undramatic stabilization of the world's population at a high material standard'.[78]

The physics professor's bright vision of the future was expressly launched as a counterweight to Ehrensvärd's diagnosis, and Gerholm himself received a lot of press coverage even before his book had reached the market. 'Finally – a prophet who does *not* preach the destruction of the world', proclaimed *Expressen*, and in *Dagens Nyheter* he was presented as an 'optimist about the future' and an alternative to the widespread 'doom-romanticism'.[79] With Gerholm's entry into the public arena, the media also began to pit Ehrensvärd and Gerholm against each other in explicit terms. *Vecko-Journalen* began an ambitiously proportioned article about the future in the following way: 'Which future do you choose? Professor Ehrensvärd's or Professor Gerholm's? With Ehrensvärd – take three steps

---

76 Tor Ragnar Gerholm, *Futurum exaktum: Fortsatt teknisk utveckling? Spekulationer om problem som måste lösas före år 2000* (Stockholm: Aldus, 1972), p. 5.
77 *Ibid.*, p. 6.
78 *Ibid.*, p. 115.
79 Clas Brunius, 'Människan överlever', *Exp*, 27 January 1972; Björn Berglund, 'Välståndet kvar spår kärnfysiker', *DN*, 27 January 1972.

backwards, scrap the car, pedal a bicycle, [and] relish the quiet charms of agrarian society. With Gerholm – continue forwards, believe in technology; but don't waste things, and ignore the doomsday prophets'.[80]

A lighthearted tone was also used by *Expressen*, which managed to set Ehrensvärd and Gerholm at each other. The meeting between the professors was featured in a double-spread article as 'the optimist versus the pessimist' and was presented as a duel in front of the blackboard. The discussion focused on four problem areas – population growth, food, water, and raw materials – but began with the professors having to mark their respective positions on a sliding scale from ultra-optimist to ultra-pessimist. 'With regard to industrial development I am a restrained optimist, whereas I place you as an ultra-pessimist', said Gerholm. Ehrensvärd agreed, but once more stressed his optimism in a long-term perspective.[81] However, Ehrensvärd's repeated attempts to modify his image continued to have difficulty taking hold in the press. It is telling that in early February *Aftonbladet* published a caricature of him in which he was described as 'the professor who became a celebrity on the basis of our downfall'.[82] Gerholm's persistent optimism about development both reinforced and clarified this media image.

Tor Ragnar Gerholm was not the only researcher to speak out against Ehrensvärd. Another was the economist Hugo Hegeland, whose debate column entitled 'Olyckskorpars låt' [The song of the ravens of doom] initiated an intense debate about the future in *Göteborgs handels- och sjöfartstidning*. Hegeland claimed that Ehrensvärd, like all other doomsday prophets throughout history, based his preaching on faith rather than on knowledge. Hegeland had been annoyed by the interview with Ehrensvärd in *Veckans affärer* in which the latter claimed to know how fast the Earth's resources were being consumed. Hegeland argued that nobody could possibly know this, because resources in the economic sense were constantly changing owing to technological development. Hegeland said that Ehrensvärd was indulging in 'mathematical

---

80 Stig Nordfeldt, 'Professorerna och framtiden', *VJ*, 9 February 1972. See also Håkan Rydén, 'Framtidsmänniskan: En bonde', *Land*, 11 February 1972.
81 Jan Lindström, 'Optimisten mot pessimisten. Hur skall det gå med människan?', *Exp*, 13 February 1972.
82 Anon., 'Gösta Ehrensvärd – professorn som blev kändis på vår undergång', *AB*, 5 February 1972.

exercises unconnected to reality'.[83] Reactions were not long in coming. One annoyed reader wrote: 'Of course, while awaiting disaster, we can stick our heads in the sand like Prof Hegeland and wait for miracles. Everything will work out fine! Sure, we might make it. But our children and grandchildren???'[84] The economist Harald Dickson felt that Hegeland was being naive in assuming that all future surprises would be favourable for humanity.[85] Hegeland responded by saying that history shows that 'when the unexpected happens, humans choose from the new possibilities those that lead to better living conditions and not to worse ones, even though we sometimes make obvious mistakes. So far, this quest has defeated the tendency towards resignation in the face of growing difficulties, as economic development overwhelmingly confirms.'[86] In the battle over the future, Hegeland – like Gerholm – believed he had history on his side. Lars Gyllensten's dilemma had arisen again.

The debate in *Göteborgs Handels- och Psjöfartstidning* gained new momentum on 10 February when Ehrensvärd responded to the criticism. He characterized Hegeland's contribution as an 'irresponsible lark song about the future' and said, as Dickson had done the day before, that it was naive to believe that the future would mainly consist of welcome surprises. This was 'not optimism but grave irresponsibility. It is simply impossible to use idle talk in explaining away the fact that we are facing major problems about humanity's development in a world of shrinking resources.' Here, and in several other places in his response, Ehrensvärd turned against the way in which optimism and pessimism were being used as paired concepts in the public debate. His most detailed piece of reasoning ran as follows:

> Planning to clean up the Earth's affairs in the long term is realism, not pessimism. No, humanity is not on its way to hell, as people love to interpret realistic warnings against unwarranted optimism. We do

---

83 Hugo Hegeland, 'Olyckskorpars låt', *GHT*, 1 February 1972. See also Karl-Göran Mäler, 'Stämmer Ehrensvärds dystra kalkyler?', *DN*, 4 February 1972.
84 Björn Kläppe, 'Hegeland har fel', *GHT*, 5 February 1972.
85 Harald Dickson, 'Blir överraskningarna angenäma?', *GHT*, 9 February 1972. See also Anon., 'Oklokt slå dövörat till för "domedagsprofeterna"', *GT*, 4 February 1972.
86 Hugo Hegeland, 'Vår framtid fångas inte gärna med en multiplikationstabell', *GHT*, 9 February 1972.

not at all have to anticipate any catastrophe (a much-loved expression) – but if matters belonging within the Western world's industrial-economic programme are allowed to run their *unrestrained* course, we may encounter unforeseen unpleasantnesses. We must therefore guard against overexploitation of the Earth's resources and take measures against overpopulation and natural destruction, *preferably right now*. A simple safety measure, nothing more. Predicting the effects of current trends is, of course, difficult in some cases, but it is not impossible to do.[87]

This passage reveals that Ehrensvärd perceived much of the ongoing debate about the future as marked by exaggerations and misunderstandings. In his view, his diagnosis was no doomsday prophecy or apocalyptic theory. It was a realistic warning that advocated taking precautions. Almost identical thoughts were expressed in Bengt Hubendick's review of *Futurum exaktum*, which was published on the same day as Ehrensvärd's reply to Hegeland. Hubendick argued that Gerholm took it for granted that future technological advances would enable humanity to finally liberate itself from its dependence on the environment. Gerholm's position, said Hubendick, would certainly be reasonable if humanity exercised control over fusion power and photosynthesis. The problem was that technological development could not be guaranteed, and its consequences could not be seen. He therefore concluded his review with 'a so-called doomsday prophet's simple question' if it would not be better to try 'to steer development towards human goals instead of advancing technological and economic development for its own sake'.[88] Gerholm did not respond to Hubendick's review.

Nor was Hubendick's piece answered by Hugo Hegeland in the latter's final contribution to the debate about the future. Instead, Hegeland addressed himself directly to Ehrensvärd and repeated his criticism of Ehrensvärd's – and other pessimists' – method of calculation. Hegeland said it was impossible, and therefore pointless, to try to predict what the world would look like in a hundred years' time. Referring to history and to the way things had turned out for other gloomy predictors of the future, he once more expressed great confidence in humanity's abilities to solve the challenges it faced. Hegeland pointed out that 'ever since the dawn of industrialization,

---

87 Gösta Ehrensvärd, 'Lärkan slår i skyn sin drill', *GHT*, 10 February 1972.
88 Bengt Hubendick, 'Kraxar den ene kvittrar den andre', *GHT*, 10 February 1972. See also Bengt Hubendick, 'Människan spår, ekosystemet rår', *GHT*, 12 February 1972.

the ravens of doom have cawed'; but time and time again, incredibly rapid economic development had exceeded even the most optimistic of expectations. He therefore firmly believed that humans – as long as they were optimistic about the future – could cope with whatever awaited them. He concluded by saying, '[i]t is my conviction that this healthy attitude will also rule among people in the future. Hence my irrepressible optimism!'[89] As these excerpts demonstrate, Hegeland vigorously confirmed the portrayal of himself as an optimist and equally consistently referred to Ehrensvärd and other 'ravens of doom' as pessimists. In other words, the polarization of the debate about the future was something that one side rejected but the other side encouraged. From the perspective of the circulation of knowledge, Ehrensvärd and Hubendick's attempts to change the parameters of the discussion had no effect.

What was plain in mid-February 1972 was thus that two clearly distinguished camps had been established in the Swedish public sphere with regard to the issue of the future: optimists and pessimists. Consequently, knowledge about an impending social collapse had been equipped with stronger reservations than before, because there were now high-profile scientists in the public arena who advocated totally different diagnoses about the future. One person who was concerned about this situation was *Svenska Dagbladet*'s Tom Selander, who argued that Gerholm and his ilk gave politicians who did not want to deal with the global problems an easy way out: 'They now get a chance to think like this: terrific, finally experts are saying that everything will work out just fine. We can calmly restrict the geographical horizon to our own constituency and the time horizon to the next election.'[90] Writing in *Dagens Nyheter*, Sven Fagerberg agreed: 'It is a damaging book that Gerholm has written. He wants to break down the responsibility that is being built up.'[91] At a later stage of the debate Fagerberg was even more censorious, arguing that *Futurum exaktum* was 'a lightly masked partisan contribution, intended to support the established industrial interests'.[92] The futurist Eskil Block also hinted that Gerholm was actually 'Wallenberg's contact man'.[93]

---

89 Hugo Hegeland, 'Pessimisternas fatala felslut!', *GHT*, 11 December 1972.
90 Tom Selander, 'En lättsinnig tänkare', *SvD*, 4 February 1972. See also Sven Gösta Nilsson, 'Tempus för framtidsforskare?', *SDS*, 4 February 1972.
91 Sven Fagerberg, 'Om Gud vill och jag får ha hälsan...', *DN*, 20 February 1972.
92 Sven Fagerberg, 'Gerholm ännu en gång', *DN*, 14 March 1972.
93 Eskil Block, 'Är framtiden vår?', *GP*, 18 February 1972.

The debaters' suspicions regarding Gerholm recall the research by historians of science Naomi Oreskes and Erik Conway into how a group of American scientists with strong ties to industry deliberately worked to create social uncertainty about scientific issues involving topics such as smoking and climate change from the 1970s onwards. The strategy depicted by Oreskes and Conway is that these researchers used their scientific authority to initiate media debates and thereby create a perception that two equal interpretations of reality opposed each other. By hinting at scientific controversies and sowing doubt about the risks to society, active measures and political interventions were hence repeatedly postponed.[94] My own research is not of such a nature that I am able to draw a conclusion about Gerholm's activities along such lines; but from a circulation perspective, it is clear that *Futurum exaktum* created – or, at least, gave a voice to – a fundamentally sceptical attitude to futures research, particularly when forecasters argued that freedom for industry and trade as well as economic growth should be curtailed. One telling example of the effects of this mode of action is found in a letter to the editor published in *Göteborgs-Tidningen*. The writer stated: 'Professors, researchers, and other scholars contradict one another when they talk about the future of the world. Some argue that the Earth will perish, others that humanity really does have a future. Who should a person believe? It is hard to decide for those of us who are less than knowledgeable.'[95]

The diagnosis as a cultural reference point

The polemical Swedish debate about the future did not end in February 1972; but after the initial exchanges in *Göteborgs handels- och sjöfartstidning*, it focused mainly on Gerholm's *Futurum exaktum* plus international agenda-setters such as Paul Ehrlich's *The Population Bomb* (1968, Swedish translation 1972), Barry Commoner's *The Closing Circle* (1972), and the Club of Rome's report *Limits to Growth* (1972).[96] In this context Gösta

---

94 Naomi Oreskes and Erik Conway, *Merchants of Doubt: How a Handful of Scientists Obscured the Truth on Issues from Tobacco Smoke to Global Warming* (New York: Bloomsbury Press, 2010).
95 M. Jansson, 'Ganska nära domedagen?', *GT*, 8 February 1972.
96 Anders Neuman, 'Okunnighet om okunnighet', *Arbt*, 15 March 1972; Anon., 'Gräns för tillväxt', *Arbt*, 3 April 1972; Leif Widén, 'Orättvist om Gerholm', *DN*, 26 February 1972; Mats Almgren et al., 'Vädjan om samarbete för

Ehrensvärd's diagnosis was repeatedly cited, but the Lund professor himself did not participate in the debate. However, his diagnosis continued to circulate in other ways; and in retrospect, Tuesday 7 March 1972 emerges as something of a turning point.

On that day Ehrensvärd had been invited to the Social Democrats' conference about the future entitled 'Är framtiden möjlig?' [Is the future possible?] in order to talk about his diagnosis and discuss the available political choices. The conference was a symbolic demonstration that the Social Democrats were taking environmental and future issues very seriously indeed, and Ehrensvärd's participation showed that his diagnosis had been deemed to possess political significance. Several of the conference participants gathered on the Tuesday evening in TV2's current affairs programme *Kvällsöppet*, which was hosted by Bo Holmström, one of Sweden's best-known heavyweights in journalism. In addition to Ehrensvärd, participants

överlevnad', *DN*, 8 March 1972; Sander, 'Marknad och människor', *DN*, 12 March 1972; Pehr G. Gyllenhammar, 'Är framtiden svensk', *GP*, 25 February 1972; Rolf Anderberg, 'Sjung hoppfullt om vinsten ändå?', *GP*, 1 March 1972; Bo Hedberg, 'Framtidsdebatten i andra andningen. Modeller av verkligheten', *GP*, 4 March 1972; Bo Hedberg, 'Modeller på framtiden = iströning för politiker', *GP*, 5 March 1972; Eskil Block, 'Om beskäftiga svenskar, cowboys och Kahn', *GP*, 6 March 1972; Peter Adler, 'Några fagra ord om vinsten', *GP*, 7 March 1972; Anon., 'Stoppad tillväxt?', *GT*, 10 March 1972; E. F., 'Glidande svar på ödesfrågor', *GT*, 14 March 1972; Björn Nihlén, 'Domedagsprofeterna har fel: Teknologin löser problemen', *HD*, 11 March 1972; Marianne Marksell, 'Katastrofen har redan börjat', *HD*, 19 March 1972; Karl-Erik Eliasson, 'Tre optimister', *HD*, 6 April 1972; Roland Pålsson, 'Ändra politiken! Rädda miljön', *Se*, 10 February 1972; Ulf Löfgren, 'Om resurser och realism', *SDS*, 11 February 1972; Nils Lewan, 'Den gröna revolutionen', *SDS*, 28 February 1972; Eskil Block, 'Olyckskorpar och korpen Bataki', *SDS*, 7 March 1972; Sven Gösta Nilsson, 'Det stationära världssamhället', *SDS*, 17 March 1972; Tor Ragnar Gerholm, 'Eskil Blocks mardrömmar', *SDS*, 21 March 1972; Mats Almgren, 'Framtiden blir inte bättre av fel sorts optimism', *SDS*, 12 April 1972; Björn Nihlén, 'Har domedagsprofeterna fel? Teknologin löser problemen', *SkD*, 20 March 1972; Hans G. Aldor, 'Katastrof eller jämvikt? Befolkningstillväxten måste stoppas före 1975', *SkD*, 29 March 1972; Harald Dickson, 'Världsdynamik – ett framtidshopp', *SvD*, 10 February 1972; Tom Selander, 'Domedagen som bestseller', *SvD*, 23 March 1972; Tom Selander, 'Priset för överlevnad', *SvD*, 27 March 1972; Tom Selander, 'Tillväxtens gränser', *SvD*, 8 April 1972; Hugo Hegeland, 'Den förstörda framtidsdebatten', *VA*, 23 March 1972; C. O. Bolang, 'Cykelsamhället – steg bakåt som leder framåt', *VA*, 30 March 1972; Eva Moberg, 'Ödets ingenjörer', *Vi*, 11 March 1972; Stig Ahlgren, 'Framför oss syndafloden', *VJ*, 22 March 1972.

in the live broadcast on 7 March included former prime minister Tage Erlander, Valfrid Paulsson of the National Environment Protection Board, and the three environmental debaters Hans Palmstierna, Björn Gillberg, and Nils-Erik Landell.

The programme's theme was environmental and future issues, and Ehrensvärd was given considerable time to talk about the energy supply of the future. Bo Holmström asked him: 'What happens to our civilization as we know it if oil and coal run out?' Ehrensvärd replied '[w]e can manage' and proceeded to outline the technological possibilities of developing nuclear breeder reactors. Even so, he emphasized that some restrictions on our material standard might be required. 'Some restrictions', Holmström wondered, 'there's talk that we'd have to go back to an agricultural environment from around the eighteenth century.' Ehrensvärd said that that might be an option if we were not to have energy from uranium: 'If we only have solar energy shining on the fields and the forest, then we will have an agrarian society again.' He then went into detail about the scientific possibilities of solving the mystery of fusion energy, claiming to be fairly optimistic about the future. Still, he did stress that it would be a different future from what we were used to. On the horizon he did not see 'the science-fiction type of a highly industrialized society', but rather a rural existence dominated by agriculture.

These remarks were followed by a conversation between Gösta Ehrensvärd and Tage Erlander about the risk of high-level political complications in the international arena. The degree of agreement between the two men was striking, and together they warned of future monopolies forming in the global energy-supply chain. For that reason, they said, research on uranium and on fusion energy should be internationalized. The debate then moved on to quality of life as a concept. Ehrensvärd argued that 'we should strive for quality instead of quantity. And quality for humans. The qualitatively living human. Not the quantity human.' At this point a sorely provoked Nils-Erik Landell interrupted and pointed out that what Ehrensvärd was talking about required far-reaching social changes, something he believed the political establishment was nowhere near trying to implement. Ehrensvärd agreed, adding that there would have to be 'a radical reorientation of our entire way of thinking. But why not? The time is ripe.'[97]

---

97 'Kvällsöppet', TV2, 7 March 1972.

The above glimpses of Ehrensvärd's television appearance show that this context was fundamentally different from the polemical press debate that had raged around him a few weeks earlier. In the medium of television, his diagnosis once again circulated as important knowledge about the future. The professor from Lund came across as a credible and thought-provoking figure of authority, respected by both the political establishment and radical environmental debaters. His optimism about the future in the fields of technology and politics was considerable.

Towards the end of March, *Sydsvenska Dagbladet* launched the reportage series 'Dina barnbarns värld' [Your grandchildren's world]. It gave a leading role to Ehrensvärd, describing *Före – Efter: En diagnos* as the starting shot for the Swedish debate about the future of humanity. Inside the newspaper, a lengthy article featured an interview with the Lund professor. The first question the reporter asked him was whether he was a doomsday prophet. Ehrensvärd said 'no', maintaining that on the contrary there was hope for humanity if austerity measures were adopted and research made progress. The reporter persisted, however, asking him to explain why his book had been labelled a doomsday prophecy. Ehrensvärd replied that status gadgets and modern comforts seemed to have become more important to humans than their continued existence. 'People today must become generationally aware in a totally different way', he claimed, adding that we should 'take the opportunity to apologize to our grandchildren for the situation in which we have placed humanity. If we can make the necessary political changes in time, then we can say that at least we realized our mistakes and changed them. Otherwise we will find it hard to look our grandchildren in the eye.'[98] The austere and gloomy picture conveyed by the interview is recognizable from previous stages of the circulation of the diagnosis, illustrating how difficult, not to say impossible, it was for Ehrensvärd to shake off the doomsday prophet epithet. Despite his persistent attempts to modify his image, he had increasingly come to personify the concept during the winter and spring of 1972. His name and his diagnosis had become cultural reference points in the social circulation of knowledge.

A telling example of this situation may be drawn from a later part of the reportage series 'Your grandchildren's world', a part which dealt with the American Amish people, who lived a pre-industrial agrarian existence for religious reasons. The journalistic

---

98 Hans Tedin, 'Dina barnbarns värld', *SDS*, 26 March 1972.

angle was 'societies like the one Gösta Ehrensvärd predicts exist today!'.[99] Another example was when the farming-orientated magazine *Land* reported that the EEC's minister for agriculture, Sicco Mansholt, had joined the debate about the future and revealed himself to be 'a regular Ehrensvärd type'.[100] A third may be taken from the cultural magazine *Ord & Bild*, which published a long essay on 'Framtidens historia: Från Jules Verne till Ehrensvärd' [The history of the future: from Jules Verne to Ehrensvärd] in the late spring of 1972.[101] The broad impact of the Lund professor's diagnosis was impossible to miss. Towards the end of May, *Land* reflected on the fears felt at the beginning of the year that politicians and builders of society would not want to concern themselves with Ehrensvärd's predictions. However, the editorial writer willingly acknowledged that *Land* had worried unnecessarily: 'A hugely extensive debate has been going on all spring about the issues made topical by Ehrensvärd. The newspapers have been full of articles. Radio and TV have done their part. And, most gratifying of all: the politicians have been paying serious attention.'[102]

Clearly, then, the ground had been well prepared for the upcoming Stockholm conference. In *Vecko-Journalen*'s big environmental issue, the magazine had not only managed to persuade Tage Erlander to lie down in the grass with two dandelions in his hand, it also boasted a specially written column by Gösta Ehrensvärd. This was the first time since 10 February that he had personally held the pen, and he used it to ask big and difficult questions from the perspective of global justice. Ehrensvärd argued that the problems required 'an array of technological expertise, wisdom, humanity, and foresight' and wondered if humankind really possessed it. The UN conference in Stockholm would supply the answer to how far we had come.[103] From my point of view, that conference was the culmination of the social breakthrough of environmental knowledge in Sweden.

---

99 Anon., 'Våra framtidsutsikter', *SDS*, 29 April 1972.
100 Anon., 'EEC skal rädda världen', *Land*, 17 March 1972.
101 Per Lysander, 'Framtidens historia. Från Jules Verne till Ehrensvärd', *Ord & Bild*, 4, 1972.
102 Anon., 'Framtiden ...', *Land*, 26 May 1972.
103 Gösta Ehrensvärd, 'Vill vi betala priset?', *VJ*, 31 May 1972.

8
A new history of knowledge

In the summer of 1971, an 11-year-old boy in Gothenburg named Mats Lidström wrote a letter to Hans Palmstierna. Mats had recently read a report on the environment in the youth magazine *Kamratposten* in which Palmstierna had participated. The report had shocked Mats. 'Is our little Tellus really in such a bad way?' he asked, adding that it was terrible that there were people who destroyed the environment just to make money. 'They should be given a real lesson' for everything they did to 'people newly born'. Now it was his generation that would be forced 'to fight against humanity's possible downfall [and for its] existence'.

To find out more about the environmental problems, Mats had bought a copy of *Plundring, svält, förgiftning*. He thought it was extremely interesting and rich in content, but also depressing. 'How can anyone be happy in this society?' he wondered. He had taken the book to school several times so that he could read a chapter out loud. Not many of his classmates had wanted to listen. 'And that is an example of why the Earth is the way it is today', he said. Personally, he was thinking of some day becoming 'one of those people who work with the environment'.[1]

Palmstierna answered Mats promptly, kindly, and at length. He agreed that money and greed all too often governed the course of events. 'Like you, I am convinced that you and others born in the 1950s and 1960s will probably have to pay dearly for the mistakes that my generation and the generations immediately before me have made.' Continued environmental destruction must be prevented. Humanity had to be protected from itself. 'In the long run, it will be a question of surviving or not, if we do not mend our ways.'

---

1 Letter from Mats Lidström to Hans Palmstierna, 17 June 1971, 452/3/6 (HP ARBARK).

A new history of knowledge

Yet it was still possible to be happy. Many forces in society were working to change the situation. And one thing was for sure, Palmstierna stressed: 'more and more people are already listening and even more will listen in future, even though your friends are not doing so right now'. In time they would change. Besides, Palmstierna told Mats that new laws were about to be enacted. Very soon it would be 'a lot harder to sell environmentally hazardous products'. There was reason to be hopeful at the international level too. Next year, the UN would arrange a big environmental conference in Stockholm. Sometimes, of course, one might feel that everything was happening too slowly and progress was too slight. But there was no reason to be discouraged. 'What you and I can do, together with many others, is to try to make the necessary changes happen sooner, so it won't be too late.'[2]

This exchange between Mats Lidström and Hans Palmstierna affords an insight into how knowledge of an environmental crisis was circulating in Swedish society in the early 1970s. It underlines the fact that knowledge was not just a matter for scientists, politicians, journalists, and environmental activists. Such knowledge could also make an 11-year-old schoolboy start to wonder. What would the world be like when he grew up? What challenges awaited him and his classmates? Was the environmental crisis really a threat to human survival?

Five years earlier, a primary schooler's worries about the future would hardly have been formulated in this way. In the mid-1960s, only a few people had talked about humanity standing on the brink of a global environmental crisis. But in the space of a few years in the late 1960s, a profound change occurred: a social breakthrough of knowledge. In this book I have mapped out and analysed how this change happened in Swedish society. It is now time for me to adopt an overall perspective. What are the most important results of this study? What does this investigation teach us about the breakthrough of environmental issues in Sweden? What new insights does this study offer to international environmental history research?

A final chapter, however, is not only a place to summarize and distil research results. It also provides an opportunity to consider the scholarly choices that were made and what their consequences were. Such consideration raises broader issues of a theoretical and methodological nature related to the study of history. What I

---

2 Letter from Hans Palmstierna to Mats Lidström, 19 June 1971, 452/3/6 (HP ARBARK).

particularly want to consider here is the history-of-knowledge approach. How does my study contribute to defining and developing the history of knowledge as a research field? What more general insights and perspectives can historians of knowledge draw from it?

Finally, I would also like to take this opportunity to look beyond the present study. Does the historical narrative I have written have any significance for us today? Does the breakthrough of environmental issues in the years around 1970 make any difference to the global challenges facing humanity in the 2020s? Can we learn anything from it? My answers to these questions will of necessity be tentative and exploratory. I am a historian, not an interpreter of modern society or a political visionary. Nevertheless, I have thought about these issues for a long time, and I have some ideas that I would like to try to put into writing. My hope is that these thoughts, inspired by the study of history, will stimulate further discussion and form a platform which provides some sort of foothold. First of all, though, I will review what I believe to be the most important results of the study.

The breakthrough of environmental issues in Sweden, 1967–1972

The first point I would like to make concerns chronology. I believe this study has established that the big breakthrough of environmental issues in a Swedish context occurred in the autumn of 1967. It was then that the environmental debate greatly intensified and its content fundamentally changed. Of course, it is possible to draw the historical lines farther back. For the history of ideas and science, the late 1940s were the turning point. But from a history-of-knowledge perspective, the autumn of 1967 is more crucial. It was not until then that knowledge of an environmental crisis seriously began to circulate in Swedish society in general. From an international perspective, this is remarkably early. Moreover, by extension via the Stockholm Conference of 1972, the Swedish breakthrough had global consequences.

My second point deals with the actors who were the driving forces behind this breakthrough. Here I would like to single out the Swedish scientists who went public to warn of a looming global catastrophe. Hans Palmstierna, Karl-Erik Fichtelius, Svante Odén, Georg Borgström and others formed a chorus of warning voices. Together they made the national environmental debate gain momentum and change direction. The Swedish researchers were part of a larger international trend of scientists becoming more openly politically committed. However, what is striking when comparing Sweden with countries of the same size is how numerous, and how relatively synchronized, the Swedish researchers were. In the neighbouring Nordic countries – which resembled Sweden in many other ways – nothing similar happened during this period.[3]

My third point has to do with the ways in which the social understanding of environmental problems changed during this breakthrough phase. It was characteristic of this period that environmental destruction began to be regarded as a matter of survival. Environmental issues thereby became linked to the two global perceptions of threat that were already circulating in society: the nuclear threat and overpopulation. These connections came to colour the rhetoric, the depictions, and the understanding of the environmental crisis in the years around 1970. From acid rain, mercury fish, and leaded petrol, connecting lines stretched out to the global level. For people like Rolf Edberg, Hans Palmstierna, Erik Isakson, and Wolter Arnberg, everything was connected. The environmental problems constituted a crisis for civilization at large.

My fourth point, however, indicates a different direction. This study has shown that the Swedish environmental debate contained other themes than the global and the apocalyptic dimensions. For actors such as Barbro Soller, Birgitta Odén, Stig Tejning, and Valfrid Paulsson, the national level had higher priority than the global one; and at the national level, the environmental debate was more low-key and focused on practicalities. It featured a profound belief that Swedish society could fix the environmental problems through a process of political decisions, ambitious research efforts, and expanded international cooperation. I would also like to emphasize that actors were able to move between, or encompass both, the two major themes of the environmental debate. The most striking example of this mobility is Hans Palmstierna.

The fifth point I would like to raise concerns the politicization of environmental issues and what this process did to the relevant knowledge. I would argue that the breakthrough phase in Sweden was characterized by relative consensus about the seriousness of the environmental problems and what needed to be done. In particular,

---

3 Jamison, Eyerman, and Cramer, *The Making of the New Environmental Consciousness*; Anker, 'Den store økologiske vekkelsen'; Räsänen, 'Converging Environmental Knowledge'; Larsson Heidenblad, 'En nordisk blick'; Notaker, 'Staging Discord'.

there was widespread agreement on the need for information and its dissemination. The key task was to raise the level of knowledge of the general public and among politicians. From 1969 onwards, however, the lines of conflict became more apparent. Politicians, trade unions, industry, and new social movements sought to lead the environmental debate in different directions. Many groups wanted to make the issues their own. One consequence of this was that the political colour of environmental issues in Sweden in the years around 1970 was unclear, or, perhaps more accurately, variable. This meant that conflicting claims to possessing knowledge were made in the public sphere, even by scientific researchers such as Gösta Ehrensvärd and Tor Ragnar Gerholm. At the time of the Stockholm Conference in June 1972, it was far from obvious what was circulating as knowledge and what was regarded as ideological position statements.

This leads on to my sixth point, which concerns the issue of power over the social circulation of knowledge. If many voices were now being raised and the substance of knowledge was becoming more and more disputed, what was it that enabled some actors to have an impact while others did not? Which knowledge arenas and social forces were most important to the circulation of knowledge? Here I would like to single out *Dagens Nyheter* as a driving factor in the Swedish environmental debate. This study has repeatedly shown how the newspaper functioned as a link between the world of research and the general public. I would also like to stress the decisive – albeit manifold – role of Swedish Social Democracy in the historical process. This role includes such things as the establishment of the government enquiry into natural resources in 1964; the establishment of the National Environment Protection Board in 1967; Folksam's 'front-against-environmental-destruction' campaign in 1968 and 1969; and ABF's study groups based on *Plundring, svält, förgiftning*. In light of this, we can also find an explanation for Hans Palmstierna's having become so influential during the late 1960s. He had access to political, media, and organizational resources that no other contemporaneous environmental debater came close to. His access is remarkable from an international perspective as well. In comparison with Palmstierna, even the most influential actors, such as Rachel Carson and Barry Commoner, were very far from actual centres of political power.

My seventh and final point concerns the emergence of the environmental movements. This grassroots involvement holds a prominent position in both the historiography and the general historical consciousness. According to this view, it was via the alternative movements that environmental awareness emerged. In a country such as Denmark, one can with some justification say that this was the case; but not in Sweden. My study reveals that the big breakthrough of environmental issues in this country occurred several years before there were any environmental movements to speak of. It was the established social forces – the world of research, the political parties, the military, the major book publishers, and the daily newspapers – that were first out of the gate. From 1969 onwards, however, new social movements and the political rituals of the New Left became increasingly important to the circulation of knowledge pertaining to the environment. In the early 1970s, Nature and Youth Sweden and Björn Gillberg's environmental groups attracted a great deal of attention with their spectacular direct actions. Like Hans Palmstierna in 1967, these actors managed to reshape the dynamics of the Swedish environmental debate. The conflicts with the established society became more forceful. The alternative movement became visible, and the concept of 'the Green Movement' [literally *gröna vågen*, the Green Wave, which encompassed a longing for the simple life in the countryside] caught on. Even so, I would argue that commitment to the environment in Sweden – both in the 1970s and today – involved significantly more people than those who were organized. Far from all of them possessed radical left-wing sympathies.

A new history of knowledge

I reached the above results by applying a history-of-knowledge approach. My ambition was to write a broad social-history narrative about the breakthrough of environmental issues in Sweden. I have done this by studying the circulation of knowledge, both in the public sphere and in the lives of individuals. Throughout the study, I have placed great emphasis on what the historical actors did and on the chronological order in which various events happened. What I have thereby wanted to show is that the social breakthrough of environmental issues was something highly concrete and human-driven: it was not an abstract cultural process that happened by itself. For that reason, too, it looked very different even within the culturally relatively uniform world of Scandinavia.[4] In order to be

---

4 Larsson Heidenblad, 'En nordisk blick'.

able to write this type of history of knowledge, I chose to focus on a relatively short period of time. This limitation made it possible for me to apply a wide-angle perspective to society.

My approach differs in fundamental ways from the previous research which has dealt in various ways with the breakthrough of environmental issues in Sweden. I want to emphasize that this research is rich and of high quality. However, the perspectives employed in it have been both broader and narrower than mine. In chronological terms, the central studies in the field have encompassed many decades. Consequently, the breakthrough years around 1970 were only a small part of a larger historical narrative. Naturally, scholars could not then construct a wide-angle perspective on society. The sole exception is Lars J. Lundgren's study of how acid rain ended up on the political agenda in 1966–1968.[5] That work, however, illustrates the second difference, which is that the previous historical narrative was narrower than mine in terms of its themes. Interest was focused on specific subjects (acid rain, criticism of growth, nuclear power), organizations (the Centre Party's Youth League, the Swedish Society for Nature Conservation, the Social Democrats), types of media (educational programmes, television journalism), and actors (Georg Borgström). In addition, there are studies of discourses and imageries in which chronology and actors did not play central parts.[6] In relation to this research, my own study is a hybrid form of empirical research and scholarly synthesis. I believe that my research both deepens and connects the existing knowledge.

---

5 Lundgren, *Acid Rain on the Agenda*; Halldén, *Demokratin utmanas*.
6 Thelander and Lundgren, *Nedräkning pågår*; Jamison, Eyerman, and Cramer, *The Making of the New Environmental Consciousness*; Bennulf, *Miljöopinionen i Sverige*; Johan Hedrén, *Miljöpolitikens natur* (Linköping: Linköping University, 1994); Anshelm, *Socialdemokraterna och miljöfrågan*; Djerf Pierre, *Gröna nyheter*; Holmberg, *Längtan till landet*; Linnér, *The World Household*; Stahre, *Den alternativa staden*; Anshelm, *Mellan frälsning och domedag*; Eva Friman, *No Limits: The 20th Century Discourse of Economic Growth* (Umeå: Department of Historical Studies, Umeå University, 2002); Anshelm, *Det vilda, det vackra och det ekologiskt hållbara*; Linnér, *Att lära för överlevnad*; Erland Mårald, *Svenska miljöbrott och miljöskandaler 1960–2000* (Hedemora: Gidlunds, 2007); Ekberg, *Mellan flykt och förändring*; Johansson, *När man skär i nuet faller framtiden ut*; Jenny Andersson and Erik Westholm, *Slaget om framtiden: Forskningens roll i konflikten mellan tillväxt och miljö* (Stockholm: Santérus, 2019); Mårald and Nordlund, 'Modern Nature for a Modern Nature'.

My characterization of the Swedish research field also holds true at the international level. There, too, the breakthrough of environmental issues tends to be studied from a longer time perspective. The focus is usually on the emergence of the environmental movements, scientific warning voices, lines of development in the history of ideas, and the level of global politics.[7] Attempts to write broader narratives of social history are rare, but they do occur. They include Michael Bess's study of French conditions and Frank Uekötter's examination of German ones.[8] The study that has the most in common with my own is Adam Rome's survey of the US Earth Day celebration on 22 April 1970. An estimated 20 million people took part in the event, and the media coverage was enormous. Around 1,500 colleges and 10,000 schools organized lectures, debates, and demonstrations. The event can be seen as a manifestation of the fact that a social breakthrough of knowledge had occurred in the United States. Earth Day channelled the growing commitment to the environment in a way that was unparalleled in the rest of the world at that time.[9] For example, the European Year of Nature Conservation in 1970 had nowhere near the same popular support and impact.

In environmental history research, the eventful years around 1970 have been characterized as 'the ecological turn'.[10] But few historians without the environmental prefix speak of that period in those terms. As Adam Rome and Frank Uekötter have pointed out, the emergence and development of modern environmental awareness is poorly integrated into the general historical narrative about the postwar period.[11] Broader narratives of social history often pass by the phenomenon itself and the relevant processes in silence. I believe that a history-of-knowledge approach which focuses on social circulation is a proactive way of trying to change this state

---

7 McCormick, *Reclaiming Paradise*; Guha, *Environmentalism*; Egan, *Barry Commoner and the Science of Survival*; Robertson, *The Malthusian Moment*; Hamblin, *Arming Mother Nature*; Rome, *The Genius of Earth Day*; Radkau, *The Age of Ecology*; Höhler, *Spaceship Earth in the Environmental Age*; Selcer, *The Postwar Origins*; Warde, Robin, and Sörlin, *The Environment*.
8 Bess, *The Light-green Society*; Uekötter, *The Greenest Nation?*
9 Rome, *The Genius of Earth Day*.
10 Engels, 'Modern Environmentalism'; Nehring, 'Genealogies of the Ecological Moment'.
11 Adam Rome, '"Give Earth a Chance"'; Frank Uekötter, 'Consigning Environmentalism to History? Remarks on the Place of the Environmental Movement in Modern History', *RCC Perspectives* 7 (2011); Uekötter, *The Greenest Nation?*, pp. 4–11.

of affairs. When environmental historians such as Adam Rome can demonstrate that they are studying very large social processes and not cultural fringe phenomena, they also gain good arguments for why their field's insights and results must be integrated into historiography at large.[12]

However, the ambition to write broad socio-historical narratives entails practical challenges, not least of an empirical nature. It is *de facto* easier to study elite actors who operated in public than to study the vast majority who did not. It is also easier to study the environmental commitment of organized activists than to access what housewives, school children, and pensioners thought and did. In the course of my research, I have repeatedly struggled with these issues. It has not been obvious how an intervention based on a history-of-knowledge methodology should be transformed into practical research. But through Hans Palmstierna's abundant correspondence I discovered people like Mats Lidström and Sören Gunnarsson, and I found others via Birgitta Odén's posthumous papers and digitally searchable newspaper archives.

Nevertheless, whether I really succeeded in my intention to study the social circulation of knowledge remains debatable. Perhaps an even larger and more varied body of empirical material would be necessary to achieve that aim? Still, I hope that this study has managed to bring out the benefits of actively striving to build up a wide-angle perspective on society. I also hope that future researchers will take over where I have left off and build on my results, for example by studying Folksam's 'front against environmental destruction' campaign, or the activities of the various educational associations, in greater detail.

Carrying out comparative and transnational projects which examine the social circulation of knowledge is at least as important. In broad syntheses, Ramachandra Guha and Joachim Radkau have shown that the breakthrough of environmental issues really was a global phenomenon.[13] What happened in Stockholm and New York in the 1970s had counterparts in Tokyo and New Delhi. When applying a global viewpoint, however, it is by no means easy to spot the individuals who made things happen in various contexts. Even very influential actors, such as Hans Palmstierna in the Swedish context, tend to become invisible. Eleven-year-old Mats Lidström

---

12 Larsson Heidenblad, 'Mapping a New History of the Ecological Turn', pp. 266, 283–284.
13 Guha, *Environmentalism*; Radkau, *The Age of Ecology*.

would scarcely be accommodated at all. But if environmental historians are serious about wanting to integrate the field within a broader historical narrative, I believe they need to use scales that lend visibility to more of the many individuals who became aware of the environmental issues. Historians also need to make concrete comparisons in order to demonstrate distinctions between the various societies.

Another hope of mine is that my own book will be able to inspire history-of-knowledge studies of other social phenomena and time periods. I am personally convinced that actor-orientated studies of the social circulation of knowledge are a fruitful way of conducting historical research. I believe there is a growing need within historical research to go beyond individual case studies and try to attain a more comprehensive grasp of key social phenomena and processes of change. I harbour no illusions that it is possible – or even desirable – to write some kind of total history of society. But within a thriving historical research field, attempts should be made to write broader social narratives, and I hope that the history of knowledge will act as an impetus for this kind of endeavour.

A historian's thoughts about the present and the future

More than half a century has passed since environmental issues made their major breakthrough in Sweden. For a historian, that is a fairly short period of time; and it is a miniscule amount of time for a climate scientist or a palaeontologist. But fifty years is a long time in a person's life. Mats Lidström's youth and most of his working life have passed. The members of Nature and Youth Sweden who demonstrated at Sergels Torg in Stockholm in 1969 are now retired. Hans Palmstierna died in a drowning accident in 1975. Birgitta Odén died in 2016 after a long and active life of scholarship. Barbro Soller passed away in January 2020.

Even so, the breakthrough of environmental issues remains a historical process that many people remember. For a few more decades, the events will continue to be contemporary history. But for the majority of people now living, including myself, the breakthrough of environmental issues in the years around 1970 is only history. It is something that happened before I was born. My knowledge of it can never be anything other than indirect. But I would still maintain that it is a form of living history. What happened then is affecting us now. The knowledge which circulated in Swedish society at that time is very similar to that circulating here today

Table 1 Similarities and differences between the ecological turn and the climate debate of our own time

Then as now

- A global perspective was applied to an interconnected set of problems.
- Hopes were placed in a circular economy.
- Researchers stressed the need for autonomy and a long-term approach.
- People sometimes expressed a sense of powerlessness about the problems.
- The problems were described as 'anthropogenic' and humans as a geological force (this was being done as early as the 1950s, that is, half a century before the concept of 'anthropocene' was coined).
- There were warnings to the effect that a blind faith in economic growth undermines the very basis of our existence.
- People stressed the urgency of steering developments in new directions.
- Some voices argued that technological solutions are not enough, and that social solutions are needed as well.
- Some people implied that democracy might have to be sacrificed in order to fix the problems.
- Campaigners believed that it was necessary to go through young people in order to reach the rest of the population.
- People said we had all the knowledge we needed – the only thing required was the will to act.
- Some voices urged that Sweden should deliberately become a global role model.
- People wondered whether we would be able to look our children and grandchildren in the eye.

Then unlike now

- At least during the breakthrough phase of 1967–1968, environmental issues were of a non-ideological nature.
- The environmental crisis was a new global survival problem. There was no fifty-year history of scientific warnings, political initiatives, and information campaigns.
- There was no organized environmental movement, and there were no green political parties.
- There was a strong trust in politicians' power and collective solutions.

(see Table 1). I am convinced that the growth and development of environmental awareness is a key part of the history of the postwar period.

However, in the half century that has passed since environmental issues made their major breakthrough, both the relevant set of problems and the knowledge about those problems have assumed different forms. Today we talk of a climate crisis rather than an environmental crisis. Of all the undesirable side effects of modern industrial society, rising temperatures are what has caused the greatest anxiety to the largest number of people in the last fifteen years. The really big problem is perhaps not that the natural resources are drying up, but the fact that they are not. Humanity might have been better served by smaller coal and oil reserves.

The 2020s began with fires in Australia and melting ice in the Arctic. Since then, the corona pandemic has overshadowed almost everything else. Many of today's children and young people therefore ask themselves the same question as Mats Lidström did in the summer of 1971: 'How is our little Tellus really doing?' The big difference is that today global mobilization is occurring at the grassroots level. By today's standards, Nature and Youth Sweden's demonstration at Sergels Torg in 1969 was small-scale. Even Earth Day 1970 pales in comparison to today's school strikes and climate marches. There was no Greta Thunberg in the late 1960s or early 1970s.

Despite these differences, as a historian I can still be surprised at how absent the medium-term perspectives of recent history are in today's environmental and climate debate. To me they are so obviously relevant. Those of us living today are not the first people to try to change society in a more sustainable direction. The Swedish government's visions of a fossil-free Sweden in 2045 are not radically different from the 1970s visions of the circular low-energy society. What would happen if an equal effort were to be invested in looking back, trying to learn from the past, as is currently being invested in simulating and forecasting the future? Please do not get me wrong. I do not believe that history has all the answers to the challenges of today and tomorrow. But nor do I think it is of no importance.

I also believe that our varying historical experiences matter more than the political debate suggests. Scientific warning voices were audible throughout the entire postwar period. Knowledge about the future has circulated in the mass media, in parliamentary bodies, and in classrooms. Many people around the world have worried about nuclear war, overpopulation, depleted natural resources, environmental destruction, nuclear power, ozone holes, and climate change. But those same people have also experienced that the future did arrive, and – so far – the great collapse of civilization as we know it has not. How do these historical experiences affect us? Do they keep us awake or lull us to sleep? What do we really believe about the future? And how do these things differ between generations and social groups?

Additional dimensions exist in the political debate. As far as Sweden is concerned, the period following my investigation came to be characterized by the existence or non-existence of nuclear power. This issue very much shook up the political landscape. In 1980 a referendum was held in Sweden. It decided that nuclear power would gradually be phased out. When I was a child in the 1990s, I learned that nuclear power plants would be shut down by 2010. This did not happen. I believe that experiences of that kind matter, too. They demonstrate the difficulties of deciding what is going to happen in thirty years' time. Within such a long time span, a society and the world around it can and will change fundamentally. It may be worthwhile to remind ourselves of that as we look ahead to 2045.

But history also gives us examples of how major and long-term changes can in fact be implemented. Environmentally hazardous substances such as DDT and CFCs have been banned and phased out. Mercury levels in Swedish lake fish have been declining for a long time. Decisions that were made in the years around 1970 have only had their full impact in our own day. These examples hint at a possible historical lesson. When the problems are concrete and delimited, the possibilities of doing something about them are good, even though implementation may take decades.

However, the really big and difficult questions are of a different nature. Fossil fuels permeate our lives and societies; DDT, CFCs, and mercury did not. Legislation and new technology may not be able to deal with all the undesirable side effects of modern industrial society. Nor is it always easy to rally round knowledge where the big and complex issues are concerned. The relative consensus that may exist during a social breakthrough of knowledge is not necessarily followed by vigorous political action. Conflicts may arise between various legitimate interests and visions.

This is also where one of the biggest differences emerges between the ecological turn and our own time. When knowledge of a global environmental crisis made its major social breakthrough in Sweden, environmental issues were not perceived in ideological terms. Representatives of various political parties and social stakeholders regarded them as scientific and technological issues which could be dealt with along rational lines. A historical window was opened which enabled an actor such as Hans Palmstierna to act as a unifying force for a time. But by the early 1970s, the dynamics had changed fundamentally. The ideological colour of the environmental issues was still not unambiguous; but there was no doubt that they were highly explosive ideologically and could cause deep and open conflicts. The Social Democratic establishment, the research community, industry and commerce, the centre-right parties, and the new environmental movements were all drawn into what was sometimes a rather messy struggle over what the concept of the environment should include and what should be done about the concomitant problems. Still, one of the few things that have been possible to agree on is the need for more research and more reliable knowledge.

Personally, however, I sometimes wonder if the environmental and climate debate is and has been excessively focused on issues to do with knowledge. It seems to me that large parts of the political debate are constructed around a naive belief that those who know what is right will do what is right. But what tells us that this is so? Experience? Science? And how do we know when something is 'right'? Perhaps the central place accorded to knowledge is a legacy from the time when environmental issues made their great global breakthrough. In many ways, the late 1960s were a culmination of the postwar belief in science and rational social planning.[14] Since then, that trust has been replaced by scepticism in many quarters.

But what if knowledge is unable to lead to a sustainable future? In a time of polarization and mistrust, should we turn our backs on science and its representatives? No, I am not saying that. What I do believe, however, is that fundamental questions about how we should live our lives – as individuals, groups, societies, and the human race – cannot be reduced to questions about science. The discussions must also be about values, principles, and historical experiences. For this reason, contemporary historical research is important to the political debate. Like all historical research, it enables us to expand our own experiential space and gain insights into what has shaped those of others. We need more of this, not less.

---

14 Francis Sejersted, *Socialdemokratins tidsålder: Sverige och Norge under 1900-talet* (Nora: Nya Doxa, 2005); Per Lundin, Niklas Stenlås, and Johan Grubbe (eds), *Science for Welfare and Warfare: Technology and State Initiative in Cold War Sweden* (Sagamore Beach, MA: Science History Publications, 2010); Åsa Lundqvist and Klaus Petersen (eds), *In Experts We Trust: Knowledge, Politics and Bureaucracy in Nordic Welfare States* (Odense: University Press of Southern Denmark, 2010); Östling, Olsen and Larsson Heidenblad (eds), *Histories of Knowledge in Postwar Scandinavia*.

Bibliography

Unpublished primary sources

Swedish Labour Movement's archives and library
The Hans Palmstierna Collection

Centre for Business History
The Aldus Förlag Archive

Department of History, Lund University
Birgitta Odén's posthumous papers

Published primary sources

Books and articles

Borgström, Georg, 'Förord', in Fairfield Osborn, *Vår plundrade planet* (Stockholm: Natur & kultur, 1949).

Carlsson, Stig, *Förbifarter* (Stockholm: Norstedts, 1968).

Carson, Rachel, *Silent Spring* (Boston, MA: Houghton Mifflin, 1962).

Dahmén, Erik, *Sätt pris på miljön: Samhällsekonomiska argument i miljöpolitiken* (Stockholm: SNS, 1968).

Edberg, Rolf, *På Jordens villkor: En trilogi om människan och hennes värld* (Stockholm: Norstedts, 1974).

Edberg, Rolf, *Spillran av ett moln: Anteckningar i färdaboken* (Stockholm: Norstedts, 1966).

Edfeldt, Åke W., *Kvicksilvergäddan* (Stockholm: Tiden, 1969).

Ehrensvärd, Gösta, *Före – Efter: En diagnos* (Stockholm: Aldus, 1971).

Fichtelius, Karl-Erik, *Människans villkor: En bok av vetenskapsmän för politiker* (Stockholm: Wahlström & Widstrand, 1967).

Bibliography

Gerholm, Tor Ragnar, *Futurum exaktum: Fortsatt teknisk utveckling? Spekulationer om problem som måste lösas före år 2000* (Stockholm: Aldus, 1972).
Gillberg, Björn O., *Hotade släktled: Genetisk bakgrund till några viktiga miljövårdsproblem* (Stockholm: Natur & kultur, 1969).
Meadows, Donella, Dennis Meadows, Jørgen Randers, and William Behrens III, *The Limits to Growth: A Report for the Club of Rome's Project on the Predicament of Mankind* (New York: New American Library, 1972).
Miljövårdsforskning. Betänkande del 1. Forskningsområdet (Stockholm: Ministry of Agriculture, 1967).
Miljövårdsforskning. Betänkande del 2. Organisation och resurser (Stockholm: Ministry of Agriculture, 1967).
Odén, Birgitta, 'Clio mellan stolarna', *Historisk tidskrift* 88 (1968).
Odén, Birgitta, 'Historiens plats i samfundsforskningen', *Statsvetenskaplig tidskrift* 71.1 (1968).
Osborn, Fairfield, *Vår plundrade planet* (Stockholm: Natur & kultur, 1949). This is a Swedish translation by Anders Byttner of Osborn's *Our Plundered Planet* of 1948.
Palmstierna, Hans, 'De klarsyntas revolt', in *Angeläget idag* (Stockholm: Sveriges Radio, 1968).
Palmstierna, Hans, *Plundring, svält, förgiftning* (Stockholm: Rabén & Sjögren, 1967).
Palmstierna, Hans with assistance from Lena Palmstierna, *Besinning* (Stockholm: Rabén & Sjögren, 1972).
Register till Riksdagens protokoll med bihang 1961–1970. Bd 3, Sakregister, L–Ö (Stockholm: Riksdagen, 1893–1971).
Sandberg, Torsten, *Framtidsmiljön: En studieplan från studieförbundet Vuxenskolan* (Falköping: Studieförbundet Vuxenskolan, 1968).
Soller, Barbro, *Nya Lort-Sverige* (Stockholm: Rabén & Sjögren, 1969).
Ward, Barbara and René Dubos, *Only One Earth: The Care and Maintenance of a Small Planet* (London: Deutsch, 1972).

Newspapers and magazines

Aftonbladet (AB)
Arbetet (Arbt)
Borlänge Tidning (BoT)
Borås Tidning (BT)
Dagens Nyheter (DN)
Expressen (Exp)
Folksam: Organ för kooperativa fackliga försäkringsrörelsen (Folksam)
Fältbiologen (FB)
Gefle Dagblad (GD)
Göteborgs Handels- och Sjöfartstidning (GHT)
Göteborgs-Tidningen (GT)

Göteborgsposten (GP)
Helsingborgs Dagblad (HD)
Hudiksvallstidningen (HT)
Jönköpingsposten (JP)
Karlstads-Tidningen (KT)
Kvällsposten (KvP)
Land
Ljungbytidningen (LT)
Norra Skåne (NS)
Norrbottenkuriren (NK)
Norrköpings Tidningar-Östergötlands Dagblad (NT-ÖD)
Ord & Bild
Se
Skaraborgs Läns Tidning (SLT)
Skånska Dagbladet (SkD)
Smålands folkblad (Sf)
Svenska Dagbladet (SvD)
Sydsvenska Dagbladet (SDS)
Tofsen
Upsala Nya Tidning (UNT)
Veckans Affärer (VA)
Veckojournalen (VJ)
Vestmanlands läns tidning (Vlt)
Vi
Värmlands folkblad (Vf)

Television and radio

'Aktuellt', SVT 27, October 1967.
'Angeläget', Sveriges Radio, 21 October 1967.
'Kvällsöppet', TV2 7, March 1972.
'Monitor', SVT 7, December 1967.

Interviews

Lars J. Lundgren, 18 September 2017.
Sverker Oredsson, 14 December 2017.

Secondary sources: books and articles

Agar, Jon, 'What Happened in the Sixties?', *British Journal for the History of Science* 41.4 (2008), 567–600.
Agar, Jon, *Science in the Twentieth Century and Beyond* (Cambridge: Polity, 2012).

Bibliography

Agrell, Wilhelm, *Svenska förintelsevapen: Utvecklingen av kemiska och nukleära stridsmedel 1928–1970* (Lund: Historiska media, 2002).
Andersson, Jenny, 'Choosing Futures: Alva Myrdal and the Construction of Swedish Future Studies 1967–1972', *International Review of Social History* 51.2 (2006), 277–295.
Andersson, Jenny, 'The Great Future Debate and the Struggle for the World', *American Historical Review* 117.5 (2012), 1411–1430.
Andersson, Jenny, *The Future of the World: Futurology, Futurists, and the Struggle for the Post-Cold War Imagination* (New York: Oxford University Press, 2018).
Andersson, Jenny and Erik Westholm, *Slaget om framtiden: Forskningens roll i konflikten mellan tillväxt och miljö* (Stockholm: Santérus, 2019).
Andersson, Jenny and Eglė Rindzevičiūtė (eds), *The Struggle for the Long-term in Transnational Science and Politics* (New York: Routledge, 2015).
Andersson, Nils and Henrik Björck (eds), *Idéhistoria i tiden: Perspektiv på ämnets identitet under sjuttiofem år* (Stockholm: Symposion, 2008).
Anker, Peder, 'Den store økologiske vekkelsen som har hjemsøkt vårt land', in John Peter Collett (ed.), *Universitetet i Oslo: 1811–2011* (Oslo: Unipub, 2011), 103–171.
Anker, Peder, *The Power of the Periphery: How Norway Became an Environmental Pioneer for the World* (Cambridge: Cambridge University Press, 2020).
Anshelm, Jonas, *Det vilda, det vackra och det ekologiskt hållbara: Om opinionsbildningen i Svenska Naturskyddsföreningens tidskrift Sveriges natur 1943–2002* (Umeå: Umeå universitet, 2004).
Anshelm, Jonas, *Mellan frälsning och domedag: Om kärnkraftens politiska idéhistoria i Sverige 1945–1999* (Eslöv: Symposion, 2000).
Anshelm, Jonas, *Socialdemokraterna och miljöfrågan: En studie av framstegstankens paradoxer* (Stockholm: Brutus Östlings Symposion, 1995).
Bennulf, Martin, *Miljöopinionen i Sverige* (Lund: Dialogos, 1994).
Berggren, Henrik, *68* (Stockholm: Max Ström, 2018).
Bergwik, Staffan, 'Kunskapshistoria: Nya insikter?', *Scandia* 84.2 (2018), 86–98.
Bergwik, Staffan, Michael Godhe, Anders Houltz, and Magnus Rodell (eds), *Svensk snillrikhet? Nationella föreställningar om entreprenörer och teknisk begåvning 1800–2000* (Lund: Nordic Academic Press, 2014).
Bergwik, Staffan and Linn Holmberg, 'Standing on Whose Shoulders? A Critical Comment on the History of Knowledge', in Johan Östling, David Larsson Heidenblad, and Anna Nilsson Hammar (eds), *Forms of Knowledge: Developing the History of Knowledge* (Lund: Nordic Academic Press, 2020), 283–299.
Berntsen, Bredo, *Grønne linjer: Natur- og miljøvernets historie i Norge* (Oslo: Unipub, 2011).
Bertilsson, Margareta, Bengt Stenlund, and Francis Sejersted, *Hinc robur et securitas? En forskningsstiftelses handel och vandel: Stiftelsen Riksbankens Jubileumsfond 1989–2003* (Hedemora: Gidlunds, 2004).

Bess, Michael, *The Light-green Society: Ecology and Technological Modernity in France, 1960–2000* (Chicago, IL: University of Chicago Press, 2003).
Bocking, Stephen, *Nature's Experts: Science, Politics, and the Environment* (New Brunswick, NJ: Rutgers University Press, 2004).
Bosbach, Franz, Jens Ivo Engels, and Fiona Watson, *Umwelt und Geschichte in Deutschland und Grossbritannien* (München: K. G. Saur, 2006).
Boström, Magnus, *Miljörörelsens mångfald* (Lund: Arkiv, 2001).
Boyer, Paul S., *By the Bomb's Early Light: American Thought and Culture at the Dawn of the Atomic Age* (Chapel Hill, NC: University of North Carolina Press, 1985/1994).
Broks, Peter, *Understanding Popular Science* (Maidenhead: Open University Press, 2006).
Burke, Peter, *What Is the History of Knowledge?* (Cambridge: Polity Press, 2016).
Collins, Harry and Robert Evans, *Rethinking Expertise* (Chicago, IL: University of Chicago Press, 2007).
Cronon, William, 'A Place for Stories: Nature, History, and Narrative', *Journal of American History* 78.4 (1992), 1347–1376.
Cronqvist, Marie, 'Bilder från nollpunkten: Visualiseringar av atomålderns urbana apokalyps', in Eva Österberg and Marie Lindstedt Cronberg (eds), *Våld: Representation och verklighet* (Lund: Nordic Academic Press, 2006), 323–342.
Cronqvist, Marie, 'Det befästa folkhemmet: Kallt krig och varm välfärd i svensk civilförsvarskultur', in Magnus Jerneck (ed.), *Fred i realpolitikens skugga* (Lund: Studentlitteratur, 2009), 169–197.
Cronqvist, Marie, 'Survival in the Welfare Cocoon: The Culture of Civil Defense in Cold War Sweden', in Annette Vowinckel, Marcus Payk, and Thomas Lindenberger (eds), *Cold War Cultures: Perspectives on Eastern and Western European Societies* (New York: Berghahn Books, 2012), 191–212.
Cronqvist, Marie, 'Utrymning i folkhemmet: Kalla kriget, välfärdsidyllen och den svenska civilförsvarskulturen 1961', *Historisk tidskrift* 128.3 (2008), 451–476.
Daston, Lorraine, 'Science, history of', in James D. Wright (ed.), *International Encyclopedia of the Social and Behavioral Sciences* (Oxford: Elsevier, 2015), 241–247.
Daston, Lorraine, 'The History of Science and the History of Knowledge', *KNOW: A Journal on the Formation of Knowledge* 1.1 (2017), 131–154.
Daum, Andreas, 'Varieties of Popular Science and the Transformation of Public Knowledge', *Isis* 100.2 (2009), 319–332.
Dauvergine, Peter, *Historical Dictionary of Environmentalism*, 2nd edition (London: Rowman & Littlefield, 2016).
Djerf Pierre, Monika, *Gröna nyheter: Miljöjournalistiken i televisionens nyhetssändningar 1961–1994* (Gothenburg: Department of Journalism, Media and Communication, University of Gothenburg, 1996).

Djerf Pierre, Monika and Lennart Weibull, *Spegla, granska, tolka: Aktualitetsjournalistik i svensk radio och TV under 1900-talet* (Stockholm: Prisma, 2001).

Doel, Ronald, 'Constituting the Postwar Earth Sciences: The Military's Influence on the Environmental Sciences in the USA after 1945', *Social Studies of Science* 33.5 (2003), 635–666.

Dunlap, Thomas, *DDT: Scientists, Citizens, and Public Policy* (Princeton, NJ: Princeton University Press, 1981).

Dupré, Sven and Geert Somsen, 'The History of Knowledge and the Future of Knowledge Societies', *Berichte zur Wissenschaftsgeschichte* 42.2–3 (2019), 186–199.

Edwards, Paul N., *A Vast Machine: Computer Models, Climate Data, and the Politics of Global Warming* (Cambridge, MA: MIT Press, 2010).

Egan, Michael, *Barry Commoner and the Science of Survival: The Remaking of American Environmentalism* (Cambridge, MA: MIT Press, 2007).

Egelston, Anne, *Sustainable Development: A History* (Dordrecht: Springer Netherlands, 2013).

Ekberg, Kristoffer, *Mellan flykt och förändring: Utopiskt platsskapande i 1970-talets alternativa miljö* (Lund: Department of History, Lund University, 2016).

Ekelund, Alexander, *Kampen om vetenskapen: Politisk och vetenskaplig formering under den svenska vänsterradikaliseringens era* (Gothenburg: Daidalos, 2017).

Ekman Jørgensen, Thomas, *Transformation and Crises: The Left and the Nation in Denmark and Sweden, 1956–1980* (New York: Berghahn, 2008).

Ekström, Anders, 'Vetenskaperna, medierna, publikerna', in Anders Ekström (ed.), *Den mediala vetenskapen* (Nora: Nya Doxa, 2004), 9–31.

Ekström, Anders (ed.), *Den mediala vetenskapen* (Nora: Nya Doxa, 2004).

Engels, Jens Ivo, 'Modern Environmentalism', in Frank Uekötter (ed.), *The Turning Points of Environmental History* (Pittsburgh, PA: Pittsburgh University Press, 2010), 119–131.

Engels, Jens Ivo, *Naturpolitik in der Bundesrepublik: Ideenwelt und politische Verhaltensstile in Naturschutz und Umweltbewegung 1950–1980* (Paderborn: Schöningh, 2006).

Engfeldt, Lars-Göran, *From Stockholm to Johannesburg and Beyond: The Evolution of the International System for Sustainable Development Governance and its Implications* (Stockholm: Government Offices of Sweden, Ministry of Foreign Affairs, 2009).

Engh, Sunniva, 'Georg Borgström and the Population–Food Dilemma: Reception and Consequences in Norwegian Public Debate, 1950s and 1960s', in Johan Östling, Niklas Olsen, and David Larsson Heidenblad (eds), *Histories of Knowledge in Postwar Scandinavia: Actors, Arenas, and Aspirations* (Abingdon: Routledge, 2020), 39–58.

Fleck, Ludwik, *Entstehung und Entwicklung einer wissenschaftlichen Tatsache: Einführung in die Lehre vom Denkstil und Denkkollektiv* (Basel: Schwabe, 1935).
Foucault, Michel, *The Essential Foucault: Selections from Essential Works of Foucault, 1954–1984* (New York: New Press, 2003).
Fox, Robert, 'Fashioning the Discipline: History of Science in the European Intellectual Tradition', *Minerva* 44.4 (2006), 410–432.
Fredrikzon, Johan, *Kretslopp av data: Miljö, befolkning, förvaltning och den tidiga digitaliseringens kulturtekniker* (Lund: Mediehistoriskt arkiv, 2021).
Frenander, Anders, *Debattens vågor: Om politisk-ideologiska frågor i efterkrigstidens svenska kulturdebatt* (Gothenburg: Department of History of Ideas and Science, University of Gothenburg, 1998).
Friman, Eva, 'Domedagsprofeter och tillväxtpredikanter – debatten om ekonomisk tillväxt och miljö i Sverige 1960–1980', *Historisk tidskrift* 121.1 (2001), 29–61.
Friman, Eva, *No Limits: The 20th Century Discourse of Economic Growth* (Umeå: Department of Historical Studies, Umeå University, 2002).
Gardeström, Elin, *Att fostra journalister: Journalistutbildningens former i Sverige 1944–1970* (Gothenburg: Daidalos, 2011).
Gedin, Per, *Förläggarliv* (Stockholm: Bonnier, 1999).
George, Alice L., *Awaiting Armageddon: How Americans Faced the Cuban Missile Crisis* (Chapel Hill, NC: University of North Carolina Press, 2003).
Gieryn, Thomas, *Cultural Boundaries of Science: Credibility on the Line* (Chicago, IL: University of Chicago Press, 1999).
Glover, Nikolas, 'Unity Exposed: The Scandinavia Pavilions at the World Exhibitions in 1967 and 1970', in Jonas Harvard and Peter Stadius (eds), *Communicating the North: Media Structures and Images in the Making of the Nordic Region* (Burlington, VT: Ashgate, 2013), 219–240.
Gogman, Lars, 'Rödgrönt samarbete med förhinder', *Arbetarhistoria* 142.2 (2012), 48.
Golinski, Jan, *Making Natural Knowledge: Constructivism and the History of Science* (Chicago, IL: University of Chicago Press, 2005).
Goodell, Rae, *The Visible Scientists* (Boston, MA: Little, Brown, 1977).
Grove, Richard, 'Environmental History', in Peter Burke (ed.), *New Perspectives on Historical Writing*, 2nd edition (Cambridge: Polity, 2001), 261–282.
Guha, Ramachandra, *Environmentalism: A Global History* (New York: Longman, 2000).
Hadenius, Stig and Lennart Weibull, *Partipressens död?* (Stockholm: Svensk informations mediecenter, 1991).
Haikola, Karl, 'Atombomben och det moderna samhället: Om framstegstankens roll i motståndet mot svensk atombomb 1956–1961' (Bachelor's degree research essay, Department of History, Lund University, 2014).

Haikola, Karl, 'Historiska perspektiv på 1970-talet', *Scandia* 86.1 (2020), 81–98.
Haikola, Karl, 'Objects, Interpretants, and Public Knowledge: The Media Reception of a Swedish Future Study', in Johan Östling, David Larsson Heidenblad, and Anna Nilsson Hammar (eds), *Forms of Knowledge: Developing the History of Knowledge* (Lund: Nordic Academic Press, 2020), 265–281.
Halldén, Daniel, *Demokratin utmanas: Almstriden och det politiska etablissemanget* (Stockholm: Department of Political Science, University of Stockholm, 2005).
Hallin, Daniel and Paolo Mancini, *Comparing Media Systems* (Cambridge: Cambridge University Press, 2004).
Hamblin, Jacob Darwin, *Arming Mother Nature: The Birth of Catastrophic Environmentalism* (Oxford: Oxford University Press, 2013).
Hammar, Isak, 'Det ständiga fallet: Romarriket som politisk resurs i samtiden', *Statsvetenskaplig tidskrift* 117.3 (2015), 451–468.
Haraldsson, Désirée, *Skydda vår natur!: Svenska naturskyddsföreningens framväxt och tidiga utveckling* (Lund: Lund University Press, 1987).
Harraway, Donna, 'Situated Knowledges: The Science Question in Feminism and the Privilege of Partial Perspective', *Feminist Studies* 14.3 (1988), 575–599.
Hedrén, Johan, *Miljöpolitikens natur* (Linköping: Linköping University, 1994).
Higuchi, Toshihiro, *Nuclear Weapons Testing and the Making of a Global Environmental Crisis* (Stanford, CA: Stanford University Press, 2020).
Holmberg, Carl, *Längtan till landet: Civilisationskritik och framtidsvisioner i 1970-talets regionalpolitiska debatt* (Gothenburg: Department of History, University of Gothenburg, 1998).
Holmberg, Gustav, 'Framtiden. Historikerna blickar framåt', in Gunnar Broberg and David Dunér (eds), *Beredd till bådadera: Lunds universitet och omvärlden* (Lund: Lund University, 2017), 280–302.
Howe, Joshua P., *Behind the Curve: Science and the Politics of Global Warming* (Seattle: University of Washington Press, 2014).
Hughes, J. Donald, *What is Environmental History?* (Cambridge: Polity, 2006).
Hünemörder, Kai F., *Die Frühgeschichte der globalen Umweltkrise und die Formierung der deutschen Umweltpolitik (1950–1973)* (Stuttgart: Steiner, 2004).
Höhler, Sabine, *Spaceship Earth in the Environmental Age, 1960–1990* (London: Pickering & Chatto, 2015).
Jamison, Andrew and Erik Baark, 'National Shades of Green: Comparing the Swedish and Danish Styles in Ecological Modernisation', *Environmental Values* 8.2 (1999), 199–218.
Jamison, Andrew, Ron Eyerman, and Jacqueline Cramer with Jeppe Læssøe, *The Making of the New Environmental Consciousness: A Comparative*

Study of the Environmental Movements in Sweden, Denmark and the Netherlands (Edinburgh: Edinburgh University Press, 1990).

Jansson, Anton, 'Things are Different Elsewhere: An Intellectual History of Intellectual History in Sweden', *Global Intellectual History* 6.1 (2021), 83–94.

Jansson, Anton and Maria Simonsen, 'Kunskapshistoria, idéhistoria och annan historia: En översikt i skandinaviskt perspektiv', *Slagmark* 81 (2020), 13–30.

Jansson, Birgitta, *Trolöshet – En studie i svensk kulturdebatt och skönlitteratur under tidigt 1960-tal* (Uppsala: Uppsala University, 1984).

Jasanoff, Sheila, 'Ordering Knowledge, Ordering Society', in Sheila Jasanoff (ed.), *States of Knowledge: The Co-production of Science and Social Order* (London: Routledge, 2004), 13–45.

Jasanoff, Sheila, Gerald E. Markle, James C. Peterson, and Trevor Pinch (eds), *Handbook of Science and Technology Studies* (Thousand Oaks, CA: SAGE, 1995).

Johansson, Gustaf, *När man skär i nuet faller framtiden ut: Den globala krisens bildvärld i Sverige under 1970-talet* (Uppsala: Uppsala University, 2018).

Jonter, Thomas, *The Key to Nuclear Restraint: The Swedish Plans to Acquire Nuclear Weapons* (London: Palgrave, 2016).

Jülich, Solveig, 'Fosterexperimentets produktiva hemlighet: Medicinsk forskning och vita lögner i 1960- och 1970-talets Sverige', *Lychnos* (2018), 10–49.

Jülich, Solveig, 'Lennart Nilsson's *A Child is Born*: The Many Lives of a Best-selling Pregnancy Advice Book', *Culture Unbound: Journal of Current Cultural Research* 7.4 (2015), 627–648.

Kaijser, Anna and David Larsson Heidenblad, 'Young Activists in Muddy Boots: Fältbiologerna and the Ecological Turn in Sweden, 1959–1974', *Scandinavian Journal of History* 43.3 (2018), 301–323.

Killingsworth, M. Jimmie and Jacqueline S. Palmer, 'Millennial Ecology: The Apocalyptic Narrative from "Silent Spring" to "Global Warming"', in Carl G. Herndl and Stuart C. Brown (eds), *Green Culture: Environmental Rhetoric in Contemporary America* (Madison, WI: University of Wisconsin Press, 1996), 21–45.

Klöfver, Helena, *Håll stövlarna leriga och för bofinkens talan: Naturintresse, miljömedvetenhet och livsstil inom organisationen Fältbiologerna* (Linköping: Tema V rapport, 1992).

Klöfver, Helena, *Miljömedvetenhet och livsstil bland organiserade ungdomar* (Linköping: Linköpings universitet, 1995).

Kraft, Alison, Holger Nehring, and Carola Sachse, 'The Pugwash Conference and the Global Cold War: Scientists, Transnational Networks, and the Complexity of Nuclear Histories', *Journal of Cold War Studies* 20.1 (2018), 4–30.

Krefting, Ellen, Espen Schaanning, and Reidar Asgaard (eds), *Grep om fortiden: Perspektiver og metoder i idéhistorie* (Oslo: Cappelen Damm Akademisk, 2017).

Kroll, Gary, 'The "Silent Springs" of Rachel Carson: Mass Media and the Origins of Modern Environmentalism', *Public Understanding of Science* 10.4 (2001), 403–420.

Kuhn, Thomas S., *The Structure of Scientific Revolutions* (Chicago, IL: University of Chicago Press, 1962).

Kupper, Patrick, '"Weltuntergangs-Vision aus dem Computer": Zur Geschichte der Studie "Die Grenzen des Wachstums" von 1972', in Frank Uekötter and Jens Hohensee (eds), *Wird Kassandra heiser? Die Geschichte falscher Ökoalarme* (Stuttgart: Steiner, 2004), 98–111.

Kärnfelt, Johan, *Allt mellan himmel och jord: Om Knut Lundmark, astronomin och den publika kunskapsbildningen* (Lund: Nordic Academic Press, 2009).

Kärnfelt, Johan, Karl Grandin, and Solveig Jülich (eds), *Kunskap i rörelse: Kungl. Vetenskapsakademien och skapandet av det moderna samhället* (Gothenburg: Makadam, 2018).

Larsson, Peter, 'Miljörörelsen', in Mats Friberg and Johan Galtung (eds), *Rörelserna* (Stockholm: Akademilitteratur, 1984), 249–263.

Larsson Heidenblad, David, 'Boken som fick oss att sluta strunta i miljön', *Svenska Dagbladet*, 23 October 2017.

Larsson Heidenblad, David, 'En nordisk blick på det moderna miljömedvetandets genombrott', in Erik Bodensten, Kajsa Brilkman, David Larsson Heidenblad, and Hanne Sanders (eds), *Nordens historiker: En vänbok till Harald Gustafsson* (Lund: Mediatryck, 2018), 113–123.

Larsson Heidenblad, David, 'Ett ekologiskt genombrott? Rolf Edbergs bok och det globala krismedvetandet i Skandinavien 1966', *Historisk tidskrift* (NO) 95.2 (2016), 245–266.

Larsson Heidenblad, David, 'Framtidskunskap i cirkulation: Gösta Ehrensvärds diagnos och den svenska framtidsdebatten, 1971–1972', *Historisk tidskrift* 135.4 (2015), 593–621.

Larsson Heidenblad, David, 'Mapping a New History of the Ecological Turn: The Circulation of Environmental Knowledge in Sweden 1967', *Environment and History* 24.2 (2018), 265–284.

Larsson Heidenblad, David, 'Miljöhumaniora på 1960-talet? Birgitta Odéns miljöhistoriska initiativ och skissernas historiografi', *Scandia* 85.1 (2019), 37–64.

Larsson Heidenblad, David, 'The Emergence of Environmental Journalism in 1960s Sweden: Methodological Reflection on Working with Digitalized Newspapers' in Johan Östling, Niklas Olsen & David Larsson Heidenblad (eds), *Histories of Knowledge in Postwar Scandinavia: Actors, Arenas, and Aspirations* (Abingdon, Oxon: Routledge, 2020), 59–73.

Larsson Heidenblad, David, 'Tillbaka till framtiden', *Statsvetenskaplig tidskrift* 118.2 (2016), 271–282.

Larsson Heidenblad, David, *Vårt eget fel: Moralisk kausalitet som tankefigur från 00-talets klimatlarm till förmoderna syndastraffsföreställningar* (Höör: Agerings, 2012).

Larsson Heidenblad, David, 'Överlevnadsdebattörerna: Hans Palmstierna, Karl-Erik Fichtelius och miljöfrågornas genombrott i 1960-talets Sverige', in Fredrik Norén and Emil Stjernholm (eds), *Efterkrigstidens samhällskontakter* (Lund: Mediehistoriskt arkiv/Media History Archives, 2019), 157–184.

Larsson Heidenblad, David and Isak Hammar, 'A Classical Tragedy in the Making: Rolf Edberg's Use of Antiquity and the Emergence of Environmentalism in Scandinavia', *International Journal of the Classical Tradition* 24.2 (2017), 219–232.

Latour, Bruno, *Reassembling the Social: An Introduction to Actor-Network-Theory* (Oxford: Oxford University Press, 2005).

Latour, Bruno, *Science in Action: How to Follow Scientists and Engineers through Society* (Cambridge, MA: Harvard University Press, 1987).

Law, John and John Hassard, *Actor Network Theory and After* (Oxford: Blackwell, 1999).

Lear, Linda, *Rachel Carson: Witness for Nature* (New York: Holt, 1997).

Lightman, Bernard, Gordon McOuat, and Larry Stewart (eds), *The Circulation of Knowledge Between Britain, India, and China: The Early-Modern World to the Twentieth Century* (Leiden: Brill, 2013).

Linnér, Björn-Ola, *Att lära för överlevnad: Utbildningsprogrammen och miljöfrågorna 1962–2002* (Lund: Arkiv, 2005).

Linnér, Björn-Ola, *The World Household: Georg Borgström and the Postwar Population-Resource Crisis* (Linköping: Tema University, 1998).

Locher, Fabien and Gregory Quenet, 'Environmental History: The Origins, Stakes and Perspectives of a New Site of Research', *Revue d'Histoire Moderne et Contemporaine* 56.4 (2009), 7–38.

Lundberg, Björn, 'The Galbraithian Moment: Affluence and Critique of Growth in Scandinavia, 1958–1972', in Johan Östling, Niklas Olsen, and David Larsson Heidenblad (eds), *Histories of Knowledge in Postwar Scandinavia: Actors, Arenas, and Aspirations* (Abingdon: Routledge, 2020), 93–110.

Lundgren, Lars J., *Acid Rain on the Agenda: A Picture of a Chain of Events in Sweden, 1966–1968* (Lund: Lund University Press, 1998).

Lundin, Per, '"Han kan bara inte hålla käften": Björn Gillberg, lantbruksvetenskapernas medialisering och 1970-talets miljödebatt' (unpublished manuscript).

Lundin, Per, *Lantbrukshögskolan och reformerna: Från utbildningsinstitut till modernt forskningsuniversitet* (Uppsala: Swedish University of Agricultural Sciences, 2017).

Lundin, Per, Niklas Stenlås, and Johan Grubbe (eds), *Science for Welfare and Warfare: Technology and State Initiative in Cold War Sweden* (Sagamore Beach, MA: Science History Publications, 2010).

Lundqvist, Lennart, *Miljövårdsförvaltning och politisk struktur* (Uppsala: Verdandi, 1971).

Lundqvist, Åsa and Klaus Petersen (eds), *In Experts We Trust: Knowledge, Politics and Bureaucracy in Nordic Welfare States* (Odense: University Press of Southern Denmark, 2010).

Lässig, Simone, 'The History of Knowledge and the Expansion of the Historical Research Agenda', *Bulletin of the German Historical Institute* 59 (2016), 29–59.

Marchand, Suzanne, 'How Much Knowledge is Worth Knowing? An American Intellectual Historian's Thoughts on the *Geschichte des Wissens*', *Berichte zur Wissenschaftsgeschichte* 42.2–3 (2019), 126–149.

Marklund, Carl, 'Double Loyalties? Small-State Solidarity and the Debates on New International Economic Order in Sweden during the Long 1970s', *Scandinavian Journal of History* 45.3 (2020), 384–406.

Markovits, Claude, Jacques Pouchepadass, and Sanjay Subrahmanyam (eds), *Society and Circulation: Mobile People and Itinerant Cultures in South Asia, 1750–1950* (London: Anthem, 2006).

Masco, Joseph, 'Bad Weather: The Time of Planetary Crisis', in Martin Holbraad and Morten Axel Pedersen (eds), *Times of Security: Ethnographies of Fear, Protest, and the Future* (New York: Routledge, 2013), 163–197.

McCormick, John, *Reclaiming Paradise: The Global Environmental Movement* (London: Belhaven Press, 1989).

McNeill, John R. and Corinna R. Unger (eds), *Environmental Histories of the Cold War* (Washington, DC: German Historical Institute, 2010).

Merton, Robert K., *On Social Structure and Science* (Chicago, IL: University of Chicago Press, 1996).

Meyer, Jan-Henrik, 'From Nature to Environment: International Organizations and Environmental Protection before Stockholm', in Wolfram Kaiser and Jan-Henrik Meyer (eds), *International Organizations and Environmental Protection* (Oxford: Berghahn, 2017), 31–73.

Müller, Simone, 'Corporate Behaviour and Ecological Disaster: Dow Chemical and the Great Lakes Mercury Crisis, 1970–1972', *Business History* 60.3 (2018), 399–422.

Mulsow, Martin and Lorraine Daston, 'History of Knowledge', in Marek Tamm and Peter Burke (eds), *Debating New Approaches to History* (London: Bloomsbury Academic, 2019), 159–187.

Mårald, Erland, *Svenska miljöbrott och miljöskandaler 1960–2000* (Hedemora: Gidlunds, 2007).

Mårald, Erland and Christer Nordlund, 'Modern Nature for a Modern Nature: An Intellectual History of Environmental Dissonances in the Swedish Welfare State', *Environment and History* 26.4 (2020), 495–520.

Nash, Roderick, 'American Environmental History: A New Teaching Frontier', *Pacific Historical Review* 41.3 (1972), 362–372.

Nehring, Holger, 'Cold War, Apocalypse, and Peaceful Atoms: Interpretations of Nuclear Energy in the British and West German Anti-Nuclear Weapons Movements, 1955–1964', *Historical Social Research/Historische Sozialforschung* 29.3 (2004), 150–170.

Nehring, Holger, 'Genealogies of the Ecological Moment: Planning, Complexity and the Emergence of "the Environment" as Politics in West Germany, 1949–1982', in Sverker Sörlin and Paul Warde, *Nature's End: History and the Environment* (Basingstoke: Palgrave Macmillan, 2009), 115–138.

Nieto-Galan, Agustí, *Science in the Public Sphere: A History of Lay Knowledge and Expertise* (Abingdon: Routledge, 2016).

Nilsson Hoadley, Anna-Greta, *Atomvapnet som partiproblem: Sveriges socialdemokratiska kvinnoförbund och frågan om svenskt atomvapen 1955–1960* (Stockholm: Almqvist & Wiksell, 1989).

Nordström, Katarina, *Trängsel i välfärdsstaten: Expertis, politik och rumslig planering i 1960- och 1970-talets Sverige* (Uppsala: Studia Historica Upsaliensia, 2018).

Notaker, Hallvard, 'Staging Discord: Nordic Corporatism in the European Conservation Year 1970', *Contemporary European History* 29.3 (2020), 309–324.

Nyhart, Lynn K., 'Historiography of the History of Science', in Bernard Lightman (ed.), *A Companion to the History of Science* (Chichester: John Wiley & Sons, 2016), 7–22.

Odén, Birgitta, 'Projektet Natur och samhälle', in Lars M. Andersson, Fabian Persson, Peter Ullgren, and Ulf Zander (eds), *På historiens slagfält: En festskrift tillägnad Sverker Oredsson* (Uppsala: Sisyfos, 2002), 315–334.

Oredsson, Sverker, *Järnvägarna och det allmänna: Svensk järnvägspolitik fram till 1890* (Lund: Rahm, 1969).

Oreskes, Naomi and Erik Conway, *Merchants of Doubt: How a Handful of Scientists Obscured the Truth on Issues from Tobacco Smoke to Global Warming* (New York: Bloomsbury Press, 2010).

Östberg, Kjell, *1968 – när allting var i rörelse: Sextiotalsradikaliseringen och de sociala rörelserna* (Stockholm: Prisma, 2002).

Östberg, Kjell and Jenny Andersson, *Sveriges historia: 1965–2012* (Stockholm: Norstedts, 2013).

Österberg, Eva, 'Birgitta Odén', in *2017 Yearbook* (Stockholm: The Royal Swedish Academy of Letters, History and Antiquities, 2017), 25–36.

Östling, Johan, 'En kunskapsarena och dess aktörer: Under strecket och kunskapscirkulation i 1960-talets offentlighet', *Historisk tidskrift* 140.1 (2020), 95–123.

Östling, Johan, 'Vad är kunskapshistoria?', *Historisk tidskrift* 135.1 (2015), 109–119.

Östling, Johan and David Larsson Heidenblad, 'Cirkulation – ett kunskapshistoriskt nyckelbegrepp', *Historisk tidskrift* 137.2 (2017), 269–284.

Östling, Johan and David Larsson Heidenblad, 'Fulfilling the Promise of the History of Knowledge: Key Approaches for the 2020s', *Journal for the History of Knowledge* 1.1 (2020), 1–6.

Östling, Johan, David Larsson Heidenblad, and Anna Nilsson Hammar, 'Developing the History of Knowledge', in Johan Östling, David Larsson Heidenblad, and Anna Nilsson Hammar (eds), *Forms of Knowledge: Developing the History of Knowledge* (Lund: Nordic Academic Press, 2020), 9–26.

Östling, Johan, David Larsson Heidenblad, Erling Sandmo, Anna Nilsson Hammar, and Kari H. Nordberg, 'The History of Knowledge and the Circulation of Knowledge: An Introduction', in Johan Östling, Erling Sandmo, David Larsson Heidenblad, Anna Nilsson Hammar, and Kari H. Nordberg (eds), *Circulation of Knowledge: Explorations in the History of Knowledge* (Lund: Nordic Academic Press, 2018), 9–33.

Östling, Johan, Niklas Olsen, and David Larsson Heidenblad, *Histories of Knowledge in Postwar Scandinavia: Actors, Arenas, and Aspirations* (Abingdon: Routledge, 2020).

Östling, Johan, Erling Sandmo, David Larsson Heidenblad, Anna Nilsson Hammar, and Kari H. Nordberg (eds), *Circulation of Knowledge: Explorations in the History of Knowledge* (Lund: Nordic Academic Press, 2018).

Paglia, Erik, 'The Swedish Initiative and the 1972 Stockholm Conference: The Decisive Role of Science Diplomacy in the Emergence of Global Environmental Governance', *Humanities and Social Sciences Communications* 8.2 (2021), 1–10.

Palmstierna-Weiss, Gunilla, *Minnets spelplats* (Stockholm: Bonnier, 2013).

Pettersson, Ingemar, *Handslaget: Svensk industriell forskningspolitik 1940–1980* (Stockholm: KTH Royal Institute of Technology, 2012).

Porter, Theodore, *Trust in Numbers: The Pursuit of Objectivity in Science and Public Life* (Princeton, NJ: Princeton University Press, 1995).

Poskett, James, 'Science in History', *Historical Journal* 63.2 (2020), 209–242.

Premfors, Rune, *Svensk forskningspolitik* (Lund: Studentlitteratur, 1986).

Radkau, Joachim, *The Age of Ecology* (Cambridge: Polity Press, 2014).

Raj, Kapil, *Relocating Modern Science: Circulation and the Construction of Knowledge in South Asia and Europe, 1650–1900* (Basingstoke: Palgrave MacMillan, 2007).

Renn, Jürgen, 'From the History of Science to the History of Knowledge – and Back', *Centaurus: An International Journal of the History of Science & its Cultural Aspects* 57.1 (2015), 37–53.

Renn, Jürgen, *The Evolution of Knowledge: Rethinking Science for the Anthropocene* (Princeton: Princeton University Press, 2020).

Roberts, Lissa (ed.), *Local Encounters and Global Circulation*, special issue of *Itinerario* 33.1 (2009).

Robertson, Thomas, *The Malthusian Moment: Global Population Growth and the Birth of American Environmentalism* (New Brunswick: Rutgers, 2012).
Rome, Adam, '"Give Earth a Chance": The Environmental Movement and the Sixties', *Journal of American History* 90.2 (2003), 525–554.
Rome, Adam, *The Genius of Earth Day: How a 1970 Teach-In Unexpectedly Made the First Green Generation* (New York: Hill and Wang, 2013).
Räsänen, Tuomas, 'Converging Environmental Knowledge: Re-evaluating the Birth of Modern Environmentalism in Finland', *Environment and History* 18.2 (2012), 159–181.
Salomon, Kim, *Rebeller i takt med tiden: FNL-rörelsen och 60-talets politiska ritualer* (Stockholm: Rabén Prisma, 1996).
Sarasin, Philipp, 'Was ist Wissensgeschichte?', *Internationales Archiv für Sozialgeschichte der deutschen Literatur* 36.1 (2011), 159–172.
Sarasin, Philipp and Andres Kilcher, 'Editorial', *Nach Feierabend: Zürcher Jahrbuch für Wissensgeschichte* 7 (2011), 7–11.
Schleper, Simone, *Planning for the Planet: Environmental Expertise and the International Union for Conservation of Nature and Natural Resources, 1960–1980* (New York: Berghahn Books, 2019).
Schneider, Ulrich Johannes, 'Wissensgeschichte, nicht Wissenschaftsgeschichte', in Axel Honneth and Martin Saar (eds), *Michel Foucault: Zwischenbilanz einer Rezeption* (Frankfurt am Main: Suhrkamp, 2003), 220–229.
Schulz, Thorsten, *Das 'Europäische Naturschutzjahr 1970': Versuch einer europaweiten Umweltkampagne* (Berlin: Wissenschaftszentrum für Sozialforschung, 2006).
Secord, James, 'Knowledge in Transit', *Isis* 95.4 (2004), 654–672.
Seefried, Elke, 'Reconfiguring the Future? Politics and Time from the 1960s to the 1980s', *Journal of Modern European History* 13.3 (2015), 306–316.
Seefried, Elke, 'Steering the Future. The Emergence of "Western" Futures Research and its Production of Expertise, 1950s to the Early 1970s', *European Journal of Futures Research* 29.2 (2014), 1–12.
Seefried, Elke, 'Towards the Limits to Growth? The Book and its Reception in West Germany and Britain 1972–1973', *German Historical Institute London Bulletin* 33.1 (2011), 3–37.
Seefried, Elke, *Zukünfte: Aufstieg und Krise der Zukunftforschung* (Berlin: De Gruyter Oldebourg, 2015).
Sejersted, Francis, *Socialdemokratins tidsålder: Sverige och Norge under 1900-talet* (Nora: Nya Doxa, 2005).
Selcer, Perrin, *The Postwar Origins of the Global Environment: How the United Nations Built Spaceship Earth* (New York: Columbia University Press, 2018).
Sellerberg, Ann-Mari, *Miljöns sociala dynamik: Om ambivalens, skepsis, utpekanden, avslöjanden m.m.* (Lund: Department of Sociology, Lund University, 1994).

Bibliography

Simonsen, Maria and Laura Skouvig, 'Videnshistorie: Nye veje i historievidenskaberne', *Temp – Tidskrift for historie* 10.19 (2020), 5–26.

Speich Chassé, Daniel and David Gugerli, 'Wissensgeschichte: Eine Standortbestimmung', *Traverse: Zeitschrift für Geschichte* 19.1 (2012), 85–100.

Stahre, Ulf, *Den alternativa staden: Stockholms stadsomvandling och byalagsrörelsen* (Stockholm: Stockholmia, 1999).

Stenfeldt, Johan, *Dystopiernas seger: Totalitarism som orienteringspunkt i efterkrigstidens svenska idédebatt* (Höör: Agerings, 2013).

Stevrin, Peter, *Den samhällsstyrda forskningen: En samhällsorganisatorisk studie av den sektoriella forskningspolitikens framväxt och tillämpning i Sverige* (Stockholm: Liber, 1978).

Svensson, Ragni, 'Pocketboken gjorde kunskapen till en konsumtionsvara', *Respons* 2 (2020).

Söderqvist, Thomas, *The Ecologists: From Merry Naturalists to Saviours of the Nation: A Sociologically Informed Narrative Survey of the Ecologization of Sweden 1895–1975* (Stockholm: Almqvist & Wiksell International, 1986).

Terrall, Mary and Kapil Raj (eds), *Circulation and Locality in Early Modern Science*, special issue of *British Journal for the History of Science* 43.4 (2010).

Thelander, Jan and Lars J. Lundgren, *Nedräkning pågår: Hur upptäcks miljöproblem? Vad händer sedan?* (Solna: National Environment Protection Board, 1989).

Thorup, Mikkel, *Hvad er idéhistorie?* (Aarhus: Slagmark forlag, 2019).

Tiberg, Joar, 'Vart tog framtiden vägen? Framtidsstudiernas uppgång och fall, 1950–1986', *Polhem: Tidskrift för teknikhistoria* 13.2 (1995), 160–175.

Topham, Jonathan, 'Rethinking the History of Science Popularization/Popular Science', in Faidra Papanelopolou, Agustí Nieto-Galan, and Enrique Perdiguero (eds), *Popularizing Science and Technology in the European Periphery 1800–2000* (Farnham: Ashgate, 2009), 1–20.

Tunlid, Anna, 'Människan och naturens överlevnad: Mottagandet av *Tyst vår* och *Plundring, svält, förgiftning* i den svenska miljödebatten' (Bachelor's degree research essay, Department of Philosophy, Lund University, 1994).

Tunlid, Anna, *Ärftlighetens gränser: Individer och institutioner i framväxten av den svenska genetiken* (Lund: Department of Cultural Sciences, History of Ideas Unit, 2004).

Tunlid, Anna and Sven Widmalm (eds), *Det forskningspolitiska laboratoriet: Förväntningar på vetenskapen 1900–2010* (Lund: Nordic Academic Press, 2016).

Uekötter, Frank, 'Consigning Environmentalism to History? Remarks on the Place of the Environmental Movement in Modern History', *RCC Perspectives* 7 (2011), 1–36.

Uekötter, Frank, *The Greenest Nation? A New History of German Environmentalism* (Cambridge, MA: MIT Press, 2014).
Uekötter, Frank (ed.), *Exploring Apocalyptica: Coming to Terms with Environmental Alarmism* (Pittsburgh, PA: Pittsburgh University Press, 2018).
Vail, David, *Chemical Lands: Pesticides, Aerial Spraying, and Health in North America's Grasslands since 1945* (Tuscaloosa, AL: University of Alabama Press, 2018).
Vogel, Jakob, 'Von der Wissenschafts- zur Wissensgeschichte der "Wissensgesellschaft"', *Geschichte und Gesellschaft* 30 (2004), 639–660.
Vogt, William, *Road to Survival* (New York: W. Sloane Associates, 1948).
Wang, Jessica, 'Scientists and the Problem of the Public in Cold War America 1945–1960', *Osiris* 17.1 (2002), 323–347.
Warde Paul, Libby Robin, and Sverker Sörlin, *The Environment: A History of the Idea* (Baltimore, MD: Johns Hopkins University Press, 2018).
Warde, Paul and Sverker Sörlin, 'Expertise for the Future: The Emergence of Environmental Prediction c.1920–1970', in Jenny Andersson and Eglė Rindzevičiūtė (eds), *The Struggle for the Long-term in Transnational Science and Politics* (New York: Routledge, 2015), 38–62.
Weibull, Lennart, 'Är partipressen död eller levande? Reflexioner från ett presshistoriskt seminarium', *Nordicom-Information* 35.1–2 (2013), 37–49.
Wennerholm (Bergwik), Staffan, *Framtidsskaparna: Vetenskapens ungdomskultur vid svenska läroverk 1930–1970* (Lund: Arkiv, 2005).
White, Richard, 'American Environmental History: The Development of a New Historical Field', *Pacific Historical Review* 54.3 (1985), 297–335.
Widmalm, Sven (ed.), *Vetenskapsbärarna: Naturvetenskapen i det Svenska samhället 1880–1950* (Hedemora: Gidlunds, 1999).
Widmalm, Sven (ed.), *Vetenskapens sociala strukturer: Sju historiska fallstudier om konflikt, samverkan och makt* (Lund: Nordic Academic Press, 2008).
Widmalm, Sven and Hjalmar Fors (eds), *Artefakter: Industrin, vetenskapen och de tekniska nätverken* (Hedemora: Gidlunds, 2004).
Wisselgren, Per, 'Vetenskap och/eller politik? Om gränsteorier och utredningsväsendets vetenskapshistoria', in Bosse Sundin and Maria Göransdotter, *Mångsysslare och gränsöverskridare: 13 uppsatser i idéhistoria* (Umeå: Umeå University, 2008), 103–119.
Wittrock, Björn, *Möjligheter och gränser: Framtidsstudier i politik och planering* (Stockholm: Liber förlag, 1980).
Wormbs, Nina, *Vem älskade Tele-X? Konflikter om satelliter i Norden 1974–1989* (Hedemora: Gidlunds, 2003).
Zelko, Frank, *Make it a Green Peace! The Rise of Countercultural Environmentalism* (New York: Oxford University Press, 2013).

Index

When names and concepts that occur in the running text also appear in footnotes, references to the latter have been omitted.

The index adheres to the Swedish alphabet, where the letters Å, Ä, and Ö follow Z.

1964 government inquiry into natural resources 19, 26–27, 29, 30, 31, 34, 43, 44, 45, 46, 48, 65, 93, 95, 106, 208

acid rain 23, 25, 27, 30–31, 41–42, 47, 130, 207, 210
Adler-Karlsson, Gunnar 125
Agar, Jon 11n20, 183n39
Agrell, Wilhelm 62n21
Ahlmark, Axel 176
Ahlqvist, Richard 135–136
Alfvén, Hannes 26
almstriden see 'Battle of the Elms'
Ander, Gunnar 128
Andersson, Jenny 4n4, 58n12, 154n27, 171n10, 186n46, 210n6
Andersson, Kristoffer 156
Andersson, Nils 11n19
Anker, Peder 4n5, 5n5, 20n44, 207n3
Anshelm, Jonas 21n47, 32, 66n39, 134n64, 145n4, n8, 164n64, 168n3, 210n6
Arnberg, Wolter 24, 123–124, 144, 152, 153, 154, 155, 158–159, 207

Baark, Erik 4n5, 20n44
Back, Pär-Erik 46–47, 93, 98, 102, 105–106, 109–110
Bartsch-Zimmer, Shadi 6n9
'Battle of the Elms' (*almstriden*, 1971) 164–165
Bengtsson, Ingemund 154–155, 160
Bengtsson, Yvonne 99
Bennulf, Martin 25n1, 143n98, 210n6
Berggren, Henrik 4n4, 154n27
Bergh, Eric 131
Berglund, Björn 21n46, 169n5, 191n67, 192n68, 194n79
Bergman, Clas 148
Bergman-Holmstrand, Ulla-Britt 114
Bergström, Sune 45, 176
Bergwik, Staffan 7n10, 12n21, 13n24, 15n29
see also Wennerholm
Berlin, Maths 177
Berntsen, Bredo 4n5, 20n44, 66n39
Bertilsson, Margareta 103n70
Bess, Michael 2n2, 183n39, 211
biocides 1, 18, 19, 25, 30, 63–66, 79, 80, 89, 114, 131, 150, 152

Björck, Henrik 11n19
Björsne, Sven-Anders 129
Block, Eskil 154, 198, 200n96
Bocking, Stephen 183n39
Bohr, Niels 61
Borgström, Georg 18, 20, 26, 34, 55–56, 63, 125, 140, 206, 210
Bosbach, Franz 92n42
Boström, Magnus 166n68
Boyer, Paul S. 60n16, 61n18, 62
Bringmark, Gösta 38–39
Brising, Lars 44
Broks, Peter 9n18
Brotaeus, Gunilla 139–140
Browaldh, Tore 44–45, 119
Burke, Peter 6, 92n42
Börjesson, Mats 140

Carlsson, Stig 132
Carson, Rachel 2, 18, 63–65, 70, 80, 99, 150, 208
 Silent Spring (1962) 2, 18, 63–65, 70, 80, 99, 150
Centre Party 142, 165, 182, 189, 210, 217
Club of Rome 4, 172, 199
 Limits to Growth (1972) 4, 172, 199
Collins, Harry 11n20
Commoner, Barry 15, 20, 118, 134–135, 199, 208
 The Closing Circle 199
Conservative Party of Sweden 99, 104–105, 133
Conway, Erik 199
Cramer, Jacqueline 4n5, 20n44, 21n47, 146n9, 207n3, 210n6
Cronon, William 92n42
Cronqvist, Marie 62n22

Dahl, Erik 64
Dahmén, Erik 44–45, 97, 98, 102, 105–107, 109–110, 140
 Sätt pris på miljön (1968) 44, 97
Dammann, Erik 20

Danielsson, Thomas 178
Daston, Lorraine 6n9, 7n10, 11n20, 15n29
Daum, Andreas 8n14
DDT 35, 63–65, 79, 90, 216
Delin, Gustaf 119
de Reus, Jacques 135
Dickson, Harald 196
Djerf Pierre, Monika 22n49, 39n37, 80n7, 83n18, 92n40, 210n6
Doel, Ronald 48n53
Dorst, Jean 15
Dubos, René 1n1
Dunlap, Thomas 18n39
Dupré, Sven 6n9

Earth Day (1970) 2n2, 18, 211, 215
Edberg, Rolf 66–70, 76, 77, 78, 85, 115, 138, 207
 Spillran av ett moln (1966) 66–70, 76–77, 85
Edfeldt, Åke W. 130
 Kvicksilvergäddan 130n54
Edwards, Paul N. 59n14
Egan, Michael 2n2, 20n45, 61n17, 135n66, 183n39, 211n7
Egelston, Anne 16n32, 17n35
Ehrensvärd, Gösta 170–172, 186–203, 208
 Före – Efter: En diagnos (1971) 170–172, 186–203
Ehrlich, Paul 20, 63, 134–135, 199
 The Population Bomb (1968) 63, 199
Einstein, Albert 61
Ekberg, Kristoffer 165n66, 210n6
Ekman Jørgensen, Thomas 4n4, 154n27
Ekström, Anders 8n14, 9n18, 183n38
Eliasson, Per 111

Index

Emmelin, Lars 122–123, 142
Engels, Jens Ivo 2n2, 15n30, 92n42, 211n10
Engfeldt, Lars-Göran 3n3, 16n32
Engh, Sunniva 18n38, 55n5
environmental journalism 22–23, 39, 78–92, 210
environmental movements 4, 17–19, 60–61, 114–167, 182–183, 208–209, 211
   Dai Dong 163
   Friends of Lake Vänern 117–118
   Friends of the Earth 16, 143, 161, 163, 165
   Greenpeace 16, 143, 160
   International Youth Federation for the Study of Conservation of Nature 162
   MIGRI, 161, 166–167
   Nature and Youth Sweden 123, 143–167, 182, 209, 213, 215
   Powwow 163
   Sierra Club, The 16
   Swedish Society for Nature Conservation 16, 55–56, 134, 145, 152, 164, 210
   Youth League of the Centre 165
Ericsson, Anders 122, 124, 138, 140, 141
Eriksson, E. (chair, 'Rädda Vänern') 117
Erlander, Tage 41–43, 46, 201, 203
European Year of Nature Conservation, The 1, 18–19, 211
Evans, Robert 11n20
Eyerman, Ron 4n5, 20n44, 21n47, 146n9, 207n3, 210n6

Fagerberg, Sven 32–33, 71–73, 76, 113, 198
Federation of Swedish Industries 168, 173, 181–182, 185

Fehrm, Martin 23, 44, 46, 47, 48, 93–94, 96–97, 98, 105, 109, 171
Fichtelius, Karl-Erik 25–26, 37, 70, 73–77, 78, 85, 118, 206
Människans villkor 26, 34–41, 67, 77, 85, 118, 120, 178
Fleck, Ludwik 11
FOA *see* Swedish National Defence Research Institute
Foerster, Svante 52
Folksam 116–117, 121–124, 127, 138, 140–141, 208, 212
Front against environmental destruction 116, 121–123, 139, 141, 208, 212
Forskningsberedningen 41–46
Foucault, Michel 11n20
Fox, Robert 11n20
Fredrikzon, Johan 59n14
Frenander, Anders 70n53
Fridell, Anders 158n46
Friman, Eva 66n39, 210n6
Fryksén, Arne 103
Fugelstad, Anders 132
Furuwidh, Karin 146
future studies 57, 93–113, 171, 186–203
Fältbiologerna see Nature and Youth Sweden

Galbraith, John Kenneth 125
Gandhi, Indira 82
Gardeström, Elin 83n19
Gedin, Per 186n48
George, Alice L. 62n23
Gerholm, Tor Ragnar 35, 132, 171–172, 193–199, 200n96, 208
Futurum Exaktum (1972) 171–172, 193–199
Gieryn, Thomas 12
Gillberg, Björn 166–167, 177–178, 183–184, 201, 209
Hotade släktled (1969) 166

global environmental crisis 1–3, 25–28, 37, 40, 46, 54–60, 66–77, 95, 163–165, 186–205, 216
Glover, Nikolas 21n47
Gogman, Lars 163n62
Golinski, Jan 11n20
Goodell, Rae 183n39
Grafström, Erik 154
Grove, Richard 92n42
Grubbe, Johan 217n14
Grut, Mario 190
Gugerli, David 5n7
Guha, Ramachandra 2n2, 16n31, 211n7, 212
Gullberg, Hans 177
Gunnarsson, Sören 23, 49–53, 117, 130, 212
Gustafsson, Lars 141
Gustavsson, Sten 176
Gyllensten, Lars 26, 35–37, 187n51, 188, 196

Hadenius, Stig 83n19
Haikola, Karl 62n21, 171n10, 186n46
Halldén, Daniel 165n65, 210n5
Hallin, Daniel 83n19
Hamati-Ataya, Inanna 6n9
Hamblin, Jacob Darwin 48n53, 54–55n1, 61n17, 211n7
Hambraeus, Gunnar 183–184
Hammar, Isak 69n48
Haraldsson, Désirée 134n64
Harraway, Donna 11n20
Hassard, John 12n23
Hasselmo, Gunilla 126n41
Hedén, Carl-Göran 35–37, 85–86, 132, 178, 186
Hedrén, Johan 210n6
Hegeland, Hugo 195–198
Heidenblad, David Larsson *see* Larsson Heidenblad, David
Helén, Gunnar 191–192

Herner, Sven 125
Higuchi, Toshihiro 61n17
history of knowledge 5–15, 204–217
 social circulation of knowledge 9–10, 14–15, 20, 23, 167, 204–217
Hjorth, Daniel 186, 193
Hofsten, Anna von 148
Holmberg, Carl 165n67, 183n37, 210n6
Holmberg, Gustav 58n12, 171n10
Holmberg, Linn 7n10, 13n24, 15n29
Holmström, Bo 200, 201
Howe, Joshua P. 48n53, 59n14
Hubendick, Bengt 33–34, 187n51, 188, 197–198
Hughes, J. Donald 92n42
Huldt, Bo 103, 105
Hultgren, Lennart 129
Hünemörder, Kai F. 2n2
Hägg, Kerstin 24, 140–141
Höhler, Sabine 2n2, 211n7

Idén, Hans 103
Incentive (business company) 174, 176, 178
Irskogen, Valfrid 133
Isakson, Erik 149–150, 154–158, 207
Ivarsson, Rune 110
Iveroth, Axel 168–170, 172, 181–182

Jacobsson, Rolf 158
Jamison, Andrew 4n5, 20n44, 21n47, 146n9, 207n3, 210n6
Jansson, Anton 11n19
Jasanoff, Sheila 11, 12n21
Johansson, Gustaf 69n48, 171n10, 210n6
Johansson, Sven 177
Jülich, Solveig 8n14, 9n18

Index

Kaijser, Anna 145n6
Kilcher, Andreas 8n12, n16
Killingsworth, M. Jimmie 60n17
Klöfver, Helena 145n7, 146n9
Kraft, Alison 61n19
Kroll, Gary 18n39, 63n26
Kronestedt, Torbjörn 159
Kroon, Sven-Åke 127
Kuhn, Thomas S. 11n20
Kupper, Patrick 172n11
Kärnfelt, Johan 9n18

Læssøe, Jeppe 4n5
Lagercrantz, Olof 71–72, 76, 120
Landell, Nils-Erik 30, 138, 155, 201
Landin, Bo 161, 162, 163
Larsson, Peter 166n68
Larsson Heidenblad, David 4–5n5, 6n9, 7n11, 8n15, 9n17, 18n38, 20n44, 22n50, 23n51, 25n1, 28n9, 39n37, 46n49, 49n57, 58n12, 62n23, 67n40, 69n48, 70n52, n54, 77n76, 80n9, 92n41, 125n39, 145n6, 170n8, 186n46, 187n50, 207n3, 209n4, 212n12, 217n14
Latour, Bruno 12
Law, John 12n23
Leander, Gun 142–143
Lear, Linda 18n39, 63n26
Liberal Party of Sweden (previously *folkpartiet*) 104, 142, 191, 192
liberal-orientated newspapers in Sweden 21, 28–29, 33–34, 38, 180, 188
Lidström, Mats 204–205, 212–213, 215
Lightman, Bernard 8n15, 11–12n20
Liljelund, Lars-Erik 154, 159n51, 160, 162

Lindbeck, Assar 46, 93, 97, 98n59, 102, 104n73, 105–106, 109
Lindblom, Paul 103, 190
Lindqvist, Lennart 124, 125n37
Linnér, Björn-Ola 18n38, 54n1, 55n5, 66n39, 210n6
Locher, Fabien 93
Lundberg, Berth 114
Lundberg, Björn 125n39
Lundgren, Lars J. 24, 25, 27n8, 99–100, 103–104, 110, 111, 112, 210
Lundholm, Bengt 34, 65, 76, 80
Lundin, Per 41n39, 166–167, 217n14
Lundqvist, Lennart 143n98
Lundqvist, Åsa 217n14
Lässig, Simone 13

Malcus, Kerstin 103
Malthus, Thomas Robert 71
Mancini, Paolo 83n19
Mansholt, Sicco 203
Marchand, Suzanne 7n10, 13
Markle, Gerald E. 11n20
Marklund, Carl 171n10
Markovits, Claude 8n15
Markos, Lauri 137
Masco, Joseph 61n17
McCormick, John 2n2, 16n31, n32, 211n7
McNeill, John R. 48n53
McOuat, Gordon 8n15
mercury 2, 20n44, 25, 27–28, 30, 35, 65–66, 79–80, 84–87, 89–90, 92, 112, 130, 153, 168, 207, 216
Merton, Robert K. 11n20
Meyer, Jan-Henrik 19n42
Michanek, Göran 34
Mijnhardt, Wijnand 6n9
Millbourn, Ingrid 99, 101
Moberg, Erik 47, 98
Moberg, Eva 187
Moberg, Rune 181

Mogensen, Harald 138
Müller, Simone 20n44
Mulsow, Martin 6n9
Myrdal, Alva 47, 171, 186
Myrdal, Gunnar 26, 125, 138
Mårald, Erland 65n35, 210n6

Naess, Arne 20
Nash, Roderick 92n42
National Environment Protection Board (now the Swedish Environmental Protection Agency) 19, 21, 26, 43–44, 48, 95–97, 99, 112, 116, 123–124, 137, 140, 153–154, 160–161, 173, 177, 201, 208
National Board for Technical Development (STU) 178
Nature and Youth Sweden (*Fältbiologerna*) 144–167, 182, 209, 215
Naturvårdsverket see National Environment Protection Board
Nehring, Holger 16n30, 60n17, 61n19, n20, 211n10
Netzén, Gösta 150
New Left 4, 17, 52, 73n64, 119, 157–165, 190–191, 209
Nieto-Galan, Agustí 9n18
Nilsson, Stig A. 81, 88, 91
Nilsson Hammar, Anna 6n9
Nilsson Hoadley, Anna Greta 62n21
Norborg, Lars-Arne 99
Nordberg, Kari H. 6n6, 8n15
Nordlund, Christer 65n35
Nordström, Katarina 153n26
Nordström, Ludvig 87–88, 90
Norrlid, Ingemar 103
Notaker, Hallvard 5n5, 19n42, n43, 207n3
nuclear power 4, 51, 62, 144, 210, 215–216
nuclear weapons 35, 39, 60–63, 66, 67–68, 70, 73, 74
Nyhart, Lynn K. 11n20

Odén, Birgitta 21, 23–24, 46–49, 78–79, 92–112, 130, 134, 140, 171, 184–186, 207, 212–213
Odén, Svante 23–24, 27, 31, 41, 47, 96–98, 152, 206
Olsen, Niklas 18n38, 217n14
Olsson, Gösta 89
Olsson, Roger 164
One World or None 61
Oppenheimer, Robert 61
Oredsson, Sverker 98–99, 101, 104, 110
Oreskes, Naomi 199
Osborn, Fairfield 17, 55–56
Our plundered planet (1948) 55
overpopulation 16, 25, 34, 46, 54–56, 58, 60, 63, 68–72, 77–78, 125, 132, 139, 186–203, 207, 215

Paglia, Erik 3n3
Palme, Olof 16–17, 126, 135, 171, 181, 191–192
Palmer, Jacqueline S. 60n17
Palmstierna, Hans 16, 22, 23, 25–53, 56, 58, 70–77, 78, 81, 85, 87, 96, 112–143, 152, 155, 167–186, 201, 204–209, 212, 215, 216
    Besinning (1972) 186
    Plundring, svält, förgiftning (1967) 21–22, 25–30, 32–34, 38, 50, 52, 67, 74, 76, 118, 120, 127–128, 134, 137, 139, 142, 179, 204, 208
Palmstierna-Weiss, Gunilla 135n69
Pauling, Linus 61
Paulsson, Valfrid 26, 96–97, 121, 154, 156, 160, 161–162, 177, 186, 201, 207
Petersen, Klaus 217n14
Peterson, James C. 11n20
Pettersson, Ingemar 182n35
Pinch, Trevor 11n20
Porter, Theodore 59n14

Index

Poskett, James 12n20
Pouchepadass, Jacques 8n15
Prebisch, Raúl 125
Premfors, Rune 41n39
Pugwash 61

Quenet, Gregory 93

Radkau, Joachim 2n2, 16n31, 17n36, 211n7, 212
Raj, Kapil 8n15
Randau, Henning 90
Reini, Marianne 156–157
Renn, Jürgen 6n7, 14
Reus, Jacques de see de Reus
Rindzevičiūtė, Eglė 22n48, 58n12
Roberts, Lissa 8n15
Robertson, Thomas 17n37, 20n45, 135n66, 211n7
Robin, Libby 2n2, 5n6, 54, 55n1, 56, 57, 58n13, 59n14, 60n15, 211n7
Rome, Adam 2n2, 18, 135n66, 211–212
Rooth, Ulf 156
Russell, Bertrand 61
Rådström, Lennart 128, 129n50
Räsänen, Tuomas 4n5, 20n44, 207n3

Sachse, Carola 61n19
Salomon, Kim 4n4, 154n27
Sandberg, Torsten 142n95
Sarasin, Philipp 5n7, 8n12
science and activism 3, 15–22, 25–41, 49–56, 61–63, 70–77, 178, 183, 186–203, 206–207
science and expertise 15, 22, 27–28, 32–49, 56–61, 64, 70–77, 165–203, 206–207
Schneider, Ulrich Johannes 5n7
Schulz, Thorsten 18n41
Secord, James 8n14, 13n26
Seefried, Elke 58n12, 172n11
Segnestam, Mats 152–153
Sejersted, Francis 103n70, 217n14

Selander, Sten 55–56
Selander, Tom 187, 198, 200n96
Selcer, Perrin 2n2, 55n1, 57n9, 211n7
Sellerberg, Ann-Mari 54n1
Simonsen, Maria 6n9, 11n19
Sjönander, Bengt 187
Skarpe, Christina 162, 163
Skouvig, Laura 6n9
SNF see Society for Nature Conservation
'So what?' (Än sen då?) 126, 136–137, 140
social democracy 19, 21, 26–27, 29, 32, 35, 38, 47, 50, 67, 87, 104–105, 116, 119, 125, 133, 138–139, 152, 155, 160, 168, 172, 174, 178–180, 182, 189–190, 200–202, 208, 210
Society for Nature Conservation, Swedish (SNF) 145, 146, 150, 151, 152, 164, 165
Soller, Barbro 21–24, 39, 76, 78–92, 111–113, 121, 129, 207, 213
  Djurfabriken (1971) 91
  Nya Lort-Sverige (1969) 22, 87–92
Somsen, Geert 6n9
Speich Chassé, Daniel 5n7
Stahre, Ulf 165n66, 210n6
Stenfeldt, Johan 70n53
Stenlund, Bengt 103n70
Stenlås, Niklas 217n14
Stevrin, Peter 41n39
Stewart, Larry 8n15
Stockholm Conference see United Nations Conference on the Human Environment
Stoltz, Larseric 128
STU see National Board for Technical Development
Subrahmanyam, Sanjay 8n15
Sundberg, Håkan 124, 126
Suominen, Teuvo 137–138
Svenningsson, Levi 186

Svensson, Ragni 186n47
Swedish National Defence
  Research Institute (FOA) 13,
  44, 46–49, 79, 93–112
Swenson, Åke 176, 183
Söderqvist, Thomas 146n9,
  151–152, 155, 157–159
Sörlin, Sverker 2n2, 5n6, 16n30,
  22n48, 28, 54, 55n1, 56, 57,
  58n13, 60n15, 61n17, 211n7

Tamm, Marek 6n9
Taube, Evert 189
Taylor, Gordon Rattray 171–172
Tejning, Stig 79–80, 84, 85, 86, 87,
  112, 176–177
Terrall, Mary 8n15
Tham, Carl 37
Thelander, Jan 25n1, 210n6
Thunberg, Greta 215
Tiberg, Joar 171n10
Tiselius, Arne 45
Toffler, Alvin 171
Topham, Jonathan 9n18
Tunlid, Anna 12n21, n22, 41n39,
  64–65

Uekötter, Frank 2n2, 15n30,
  18–19, 61n17, 172n11, 211
Ulfstrand, Staffan 29–30
United Nations 1–4, 125
  conference on the human
    environment 1–4, 16–17,
    162–163, 167, 171–172, 203,
    205, 206, 208
  conference on trade and
    development (UNCTAD) 125
Unger, Corinna R. 48n53

Vail, David 18n39
Vattenfall 134, 154, 156, 164
Vestberg, Ragnar 79, 80
Vindel river 144, 152
Vogel, Jakob 5n7
Vogel, Viveka 126

Vogt, William 17, 55–56
  Road to Survival (1948) 55–56
von Hofsten, Anna *see* Hofsten,
  Anna von
Vänern (Lake) 27, 84, 86, 117–118
Väröbacka 51, 152

Wang, Jessica 61n18
Wanhainen, Väinö 154
Ward, Barbara 1n1
Warde, Paul 2n2, 5n6, 16n30,
  22n48, 28, 54, 55n1, 56, 57,
  58n13, 59n14, 60n15, 61n17,
  211n7
Watson, Fiona 92–93n42
Weibull, Lennart 83n18
Wennerholm, Staffan (*see also*
  Bergwik) 146
Westholm, Erik 210n6
Westrup, Zenon P. 133–134
White, Richard 92n42
Wiberger, Ingvar 183
Wickman, Krister 35, 120
Widmalm, Sven 11n20, 12n21,
  n23, 41n39
Widmark, Gunnar 79
Wigforss, Ernst 105
William-Olsson, H. 136–137
Wisselgren, Per 12n22
Wittrock, Björn 171n10
Wormbs, Nina 12n23
Wrenfelt, Birgitta *and* Bo 120
Wretlind, Arvid 84

Zelko, Frank 16n31, 160n54

Åse, Lars-Erik 146

Än sen då? see 'So what?'

Öhman, Ivar 136
Östberg, Kjell 4n4, 154n27
Österberg, Eva 78n2
Östling, Johan 6, 7n10, n11, 8–10,
  13, 56n6, 217n14

EU authorised representative for GPSR:
Easy Access System Europe, Mustamäe tee 50,
10621 Tallinn, Estonia
gpsr.requests@easproject.com

www.ingramcontent.com/pod-product-compliance
Ingram Content Group UK Ltd.
Pitfield, Milton Keynes, MK11 3LW, UK
UKHW021823140426
5217IPUK00004B/71